THE ELEMENTS OF THE INTEGRAL CALCULUS

With its Applications to Geometry and to the Summation of Infinite Series

MAVEN BOOKS

THE ELEMENTS OF THE INTEGRAL CALCULUS

With its Applications to Geometry and to the Summation of Infinite Series

J.R. Young

MAVEN BOOKS

Chennai   Tirchy   Tirunelveli   New Delhi

MAVEN BOOKS

An Imprint of **MJP Publishers**

ISBN 978-93-88191-03-6 **Maven Books**

All rights reserved   No. 44, Nallathambi Street,
Printed and bound in India   Triplicane, Chennai 600 005

MJP 566   © Publishers, 2018

Publisher : C. Janarthanan

Publisher's Note

The legacy of a country is in its varied cultural heritage, historical literature, developments in the field of economy and science. The top nations in the world are competing in the field of science, economy and literature. This vast legacy has to be conserved and documented so that it can be bestowed to the future generation. The knowledge of this legacy is slowly getting perished in the present generation due to lack of documentation.

Keeping this in mind, the concern with retrospective acquiring of rare books has been accented recently by the burgeoning reprint industry. Maven Books is gratified to retrieve the rare collections with a view to bring back those books that were landmarks in their time.

In this effort, a series of rare books would be republished under the banner, "Maven Books". The books in the reprint series have been carefully selected for their contemporary usefulness as well as their historical importance within the intellectual. We reconstruct the book with slight enhancements made for better presentation, without affecting the contents of the original edition.

Most of the works selected for republishing covers a huge range of subjects, from history to anthropology. We believe this reprint edition will be a service to the numerous researchers and practitioners active in this fascinating field. We allow readers to experience the wonder of peering into a scholarly work of the highest order and seminal significance.

Maven Books

PREFACE.

THE work here submitted to the notice of the public forms the third volume of a course intended to furnish to the mathematical student a pretty comprehensive view of the principles of modern analytical science. To complete this design will require a fourth volume, in some measure supplementary to the three now completed, and to contain the subject of Finite Differences, a fuller inquiry into the theory of Partial Differential Equations, and a chapter on Definite Integrals. This final volume I hope hereafter to be able to prepare, although I do not propose to enter immediately upon the undertaking.

With respect to this third volume, I ought to observe that, in common with all modern elementary writers, I have availed myself pretty freely of the writings of the French mathematicians. In stating this, I am aware that I am not offering any apology for my book; but am, on the contrary, setting forth its principal claim to the notice of the English student; for the superiority of the French in every department of abstract science, is now pretty generally acknowledged in this country. Notwithstanding this admission, however, I have long been persuaded that many of the French processes, now universally adopted in English Books, are very deficient in mathematical rigour, and in not a few cases fail altogether to establish the conclusions aimed at. In consequence of this conviction, I have therefore been led, in preparing these volumes, cautiously to examine whatever I have appropriated from the sources referred to, and the result has been, that objections of the gravest kind have been found to attach to some of the most celebrated French theories. In science, as in morals, the propagation of error is of more dangerous tendency than the suppression of truth; and if, in the course of these volumes, it be found that I have succeeded in removing any inaccuracies that may hitherto have vitiated the purity of mathematical reasoning, it may perhaps atone, in some measure, for the absence of that kind of originality which requires powers of altogether a higher order.

The present treatise I have divided into three sections: the first being devoted to the Integration of Differentials of One Variable; the second to the subjects of Rectification, Quadrature, and Cubature; and the third to an Elementary View of the Theory of Differential Equations, more particularly those of the First and Second Order.

The first section will be found to be tolerably extensive. I have endeavored to arrange the several topics it embraces, so as to facilitate the progress of the student, and with the same view I have, in some cases, presented the general formulas of integration in a tabular, and I think somewhat improved, form. The sixth chapter of this section, which treats on the Methods of Integrating by Series, and on Successive Integration, will I believe, be found to contain one or two facilitating processes worth the student's attention; also in the following, or seventh chapter, the article on the Summation of Series will, it is hoped, be acceptable to the young analyst. This is a department of pure mathematics of considerable importance, as well as difficulty, and one to which the Integral Calculus is peculiarly applicable, although, in general, but a very inadequate space is allotted to it in books on that subject. In the course of this chapter occasion is taken to introduce *Wallis's* remarkable expression for the quadrant of a circle; this expression is very generally known among mathematicians, and in foreign books is always given correctly. In all the recent English works, however, which I have seen, and in which this expression occurs, it is transformed into an absolute absurdity; for in some of these books Wallis is made to say, and the student gravely informed, that the circumference of a circle whose radius is unity is accurately *nothing*; and in others the expression tells us that the circumference of the same circle is *infinite!*

The second section may be considered as the geometrical application of the first, and will be found to contain a very copious collection of problems on Rectification, Quadrature, and Cubature; most of these problems have been selected from different mathematical periodicals, but of the greater part of these the solutions have been modified and improved, and corrected where erroneous.

It may be objected that I have not introduced into this section the usual ancient curves, as the Quadratrix, the Conchoid, the Cissoid, &c., the truth is that I think too much attention is bestowed on these curves at the present day, as they have long been dispossessed of that interest and importance that attached to them at the time of their invention; and, moreover, in the present improved and extended state of mathematical science, an ordinary student will find it a matter sufficiently difficult to preserve in his memory the many particulars which it is of importance should be remembered, without being burthened, in addition, with the names and forms of the various curves devised by the early geometers in their fruitless attempts to square the circle, to trisect an angle, and to double a cube. On these accounts I have not hesitated to exclude them from this treatise and to introduce others, offering, by their equations, more interesting analytical particulars.

a 2

The third section contains the elements of a theory of almost boundless extent, the theory of Differential Equations. As far as equations of two variables and of the first order are concerned, and beyond which the powers of the calculus are at present but very limited, the information conveyed in this section will, it is thought, be found to be sufficiently copious. I have endeavoured to render this part of the subject clear and intelligible, and have, in some cases, preferred appearing lengthy where brevity might involve any obscurity, as in the article on *Riccati's equation*, for instance In the latter part of this section I have compressed into small compass several topics of a nature too difficult and too extensive to be completely discussed in a work of this kind; but I have taken care to direct the inquiring student to the sources where more satisfactory information may be obtained: I hope, also, to touch again upon these matters at a more convenient opportunity; in the mean time, I trust that the three volumes now finished may contribute something towards improving the taste and exciting the inquiries of the young analyst.

J. R. YOUNG.

August 25, 1831.

CONTENTS.

SECTION I.

On the Integration of Differential Expressions of a Single Variable.

Article   *Page*

1. Integration defined 1
2. Symbol of Integration . . . . . . ib.
3. Obvious particulars respecting the fundamental principles of integration . . . . ib.
4. Integration of the form $(Fx)^n d\,Fx$ 2
5. Examples of this form 3
6. Particular case of the above form requiring separate consideration 6
7. Integration of the forms

$$+ \int \frac{dx}{\sqrt{a^2 - b^2 x^2}} \,,\; \pm \int \frac{dx}{a^2 + b^2 x^2} \,,\; \pm \int \frac{dx}{x\sqrt{b^2 x^2 - a^2}} \,,\; \pm$$

$$\int \frac{dx}{\sqrt{a^2\, x - b^2 x^2}} \qquad . \qquad . \qquad . \quad 8$$

8. Integration of these forms when $a = b = 1$, &c. . . 9
9. Expressions in art. 7 modified so as to require only trigonometrical
tables . . . . . . ib.
10. Examples of circular forms 10
11. Elementary exponential and trigonometrical forms . . 12
12. Integration of rational fractions 13
13. Case in which imaginary factors enter the denominator . 22
14. Examples in this case 25
15. Reduction of irrational functions to rational . . . 28
16. When binomial surds enter the function . . . 29
17. When the surds are all of the form $\sqrt{a + bx + cx^2}$ . . 31
18. Examples in this case 32
19. On the integration of binomial differentials . . . 35

Article   *Page*

20. Condition of integrability of the general form

$$x^m (a + bx^n)^p \, dx \qquad . \qquad . \qquad . \quad 36$$

21. Another condition of integrability   37
22. Examples   .   .   .   . ib.
23. Methods of reduction when the conditions of integrability are not
    satisfied   39
24. Table of formulas for the reduction of the integral

$$x^m (a + bx^n)^p \text{ or } \int x^m \, {}_z{}^p \, dx \quad . \qquad . \quad . \quad 42$$

25. Examples of their application   43
26. On the integration of logarithmic and exponential forms   . 52
— To integrate the form X log. $^n x dx$.   .   .   . ib.
27. To integrate the less general form x^m log.$^n x dx$.   . 53
28. Examples   55
29. The formula for the differential in art. 27 applicable also to the
    form $(x \pm a)^m \, dx$ log.$^n x$ 56
30. To integrate the form $a^x x^m \, dx$ 57
31. Examples   58
32. Integration of trigonometrical and circular functions   .   . 60
33. To integrate the form sin.^{m} cos.$^n x dx$.   .   .   . ib.
34. General formulas for the reduction of the integral

$$\int \sin.^m x \, \cos.^n x \, dx \qquad . \qquad . \quad . \quad 62$$

35. Examples of their application   64
36. To integrate the forms

$$\frac{\sin.^m x}{\cos.^n x} \, dx, \quad \frac{\cos.^n x}{\sin.^m x} \, dx \qquad . \qquad . \quad 65$$

37. General formulas for the integrals

$$\int \frac{dx}{\sin.^m x} \text{ and } \int \frac{\cos.^n x}{\cos.^n x} \qquad . \quad 68$$

38. Examples of their application   70
39. Integration of the cases in which m and n are positive whole
    numbers in terms of simple powers of the trigonometrical lines 71
40. To integrate the form

$$\frac{dx}{\sin.^m x \, \cos.^n x} \qquad . \qquad . \qquad . \quad 73$$

41. Examples of this form   .   .   .   .   .   . ib.

Article   *Page*

42. Cases in which the form $\sin.^m x \cos.^n x dx$ is integral for fractional values of m and n 75

43. To integrate the forms

$$m^{ax} \sin.^n x\, dx \text{ and } m^{ax} \cos.^n x\, dx \qquad . \qquad . \quad 76$$

44. Examples of this form 77

45. To integrate the forms

$$X \sin.^{-1} x\, dx \text{ and } X \cos.^{-1} dx \qquad . \qquad . \quad 78$$

46. Examples of these forms 79
47. On Integration by series and on successive integration . . 80
48. The series of *John Bernouilli* 81
49. Another series derived from *Maclaurin's theorem* . . ib.
50. Another mode of obtaining the series for $\int X dx$ . . ib.
51. This last mode leads to a method of facilitating development . 82
52. Examples of integration by series 83
53. Successive integration 89
54. Series for the development of $\int^n X dx$ 90
55. Examples 91
56. Integration between limits 94
57. Examples 95
58. Summation of Series 99

SECTION II.

On Rectification, Quadrature, and Cubature.

59. Formula for the length of a plane curve . . . 112
60. Rectification of parabolas and of the circle . . . 113
61. Rectification of the ellipse 115
62. Rectification of the cycloid and various other curves . 119
63. Formula to be employed when the curve is referred to polar coordinates 123
64. Another expression for the curve 124
65. Rectification of various polar curves . . . . ib.
66. Formula for the quadrature of a curve . . . 126
67. Application of the formula to a variety of curves . . 127
68. Expressions for the integrals of $\sqrt{r^2 - x^2}\, dx$ and $\sqrt{2rx - x^2}\, dx$ by means of circular segments 128

x.

CONTENTS.

Article   *Page*

69. Formula for the quadrature of polar curves   136
70. Application to various polar curves   137
71. Formula for the quadrature of curve surfaces   .   .   . 142
72. Application to examples   144
73. On cubature   147
74. Cubature of volumes generated by the parallel motion of a plane   150
75. Formulas for the quadrature of surfaces, and for the cubature of
      volumes referred to three coordinate planes   .   . 153
76. Miscellaneous processes of integration   .   .   . 159

SECTION III.

On the Integration of Differential Expressions of several Variables.

77. Integration of exact differentials   170
78. *Euler's* criterion of integrability   171
79. Differentials which satisfy the criterion of integrability   . 173
80. Differentials which are both exact and homogeneous   .   . 178
81. Examples   180
82. On the theory of differential equation and of arbitrary constants   182
83. Illustration of this theory by examples   .   .   . 188
84. On the integration of differential equations of two variables of
      the first order and degree   190
85. Separation of the variables   .   .   .   .   . ib.
86. Separation of the variables in the form

$$XY\,dy + X'Y'\,dx = 0 \quad . \qquad . \qquad . \quad 191$$

87. Separation of the variables in the linear equation

$$dy + Xy\,dx = X'\,dx \quad . \qquad . \qquad . \quad ib.$$

88. Series summable by means of linear differential equations   . 194
89. The separation of the variables always possible in differential
      equations of the first order whenever they are homogeneous   197
90. The *Equation of Riccati*   202
91. Examples of Riccati's equation   207
92. Methods of rendering differentials exact   .   .   . 210
93. Determination of the requisite factor when it is a function of
      but one of the variables   211

Article   *Page*

91. Determination of the factor in homogeneous differential equations 211
95. To differentiate under the sign of integration . . . 217
96. Geometrical applications of the Calculus . . . 218
— On Trajectories 220
97. Integration of differential equations of the first order and of the higher degrees 221
98. Examples 225
99. The theory of singular solutions 232
100. Criteria of singular solutions 235
101. Geometrical interpretation of singular solutions . . 237
102. Examples of the determination of singular solutions . . 238
103. Geometrical problems conducting to singular solutions . 240
104. Integration of differential equations of the second and higher orders 243
105. Integration of certain cases by indirect processes . . 251
106. Certain integrable forms of equations of a higher order than the second 253
107. Examples of these forms 254
108. Linear equations of the higher orders . . . 256
109. Determination of integrals by approximation . . 259
110. Integration of simultaneous equations . . . 264
ih. Integration of total differential equations of three variables . 268
111. Example of the case which satisfies the condition of integrability 271
112. Case in which the condition is not satisfied . . 272
113. Integration of partial differential equations of the first order 274
114. Examples of partial differential equations . . . 276
115. On the determination of arbitrary functions . . 283
NOTES 289-306

ERRATA.

Page 2, line 4   from top   *for* $\int a\mathrm{F}'dx$, *read* $\int a\mathrm{F}'xdx$.

9, ... 3   ...   ... $\dfrac{a}{2b}\sqrt{2}$ *read* $\dfrac{a^2}{2b^2}$.

24, ... 7   ...   ... $\displaystyle\int \dfrac{x^2 dz}{(z^2+\beta^2)^m}$ *read* $\displaystyle\int \dfrac{z^2 dz}{(z^2+\beta^2)^m}$.

35, .. 8 from bottom, ... $x^m(a+bx^n)^{\frac{p}{q}}$ *read* $x^m(a+bx^n)^p$.

42, ... 6   ..   ... $\int z^m z^p\, dx$, *read* $\int x^m z^p\, dz$.

90, ... 2   from top   ... $dn^{-1} y$, *read* $d^{n-1} y$.

106, ... 6 from bottom, ... $+$ &c. *read* $-$ &c.

127, ... 9   ...   ... arc, *read* area.

209, ... 7   from top   ... he, *read* the.

214, ... 11 from bottom, ... $d'u$, *read* du'.

THE

INTEGRAL CALCULUS.

SECTION I.

ON THE

INTEGRATION OF DIFFERENTIAL EXPRESSIONS OF A SINGLE VARIABLE.

Article (1). THE Integral Calculus is the reverse of the Differential Calculus, its object being to determine the primitive function from which any proposed differential is derived. We shall at once proceed to the exposition of the principles of this very important department of Analysis.

CHAPTER I.

FUNDAMENTAL PRINCIPLES OF INTEGRATION.

(2.) The process by which we return from the derived function to the primitive is called integration; it is indicated by the symbol $\int$ placed before the differential or derived function, and the result of the process, that is, the primitive function, is called the integral of the proposed differential.

(3.) There are several obvious particulars respecting the fundamental principles of integration, which immediately present themselves to the mind, from considering the direct process, or that of differentiation. These we shall briefly enumerate:

1. Since $da\mathrm{F}x$ is the same as $ad\mathrm{F}x$, viz. $a\mathrm{F}'xdx$, therefore, in the reverse operation, $\int a\mathrm{F}'dx$ is the same as $a\int \mathrm{F}'xdx$, so that any constant factor or divisor may be taken from under the sign of integration, and placed without it. We may moreover introduce any constant factor under the sign $\int$, provided we place its reciprocal without the sign.

2. Since the differential of the sum of any number of functions is the same as the sum of their several differentials, it follows that, when we have to integrate the sum of any number of differentials, the same integral will be expressed, whether the sign $\int$ is prefixed to the whole sum, or to each individual differential, that is, $\int(\mathrm{A}dx + \mathrm{B}dx + \&c.)$ is the same as $\int \mathrm{A}dx + \int \mathrm{B}dx + \&c.$

3. Since, in differentiating any function, the constant connected with it by addition or subtraction disappears from the result, it follows that, in integrating such result, the constant should be introduced. But as the form of the differential remains the same, whatever may have been the constant in the primitive, we cannot infer from that form the particular value of the constant that has disappeared, so that all we can do is, to annex to the integral found a symbol C, standing for a constant, the value being indeterminate. The integral thus completed has the most general form possible, since it comprehends every function that can by differentiation produce the proposed differential. Thus the complete integral of the differential $d\mathrm{F}x$ is $\mathrm{F}x + \mathrm{C}$. If we know in any particular inquiry what value the integral ought to take for any one particular value of the variable, the constant belonging to that case becomes readily determinable. Thus, if we know that for $x = a$ the value of the integral $\mathrm{F}x + \mathrm{C}$ ought to be A, then we have $\mathrm{F}a + \mathrm{C} = \mathrm{A}$, therefore the value of the constant is in that case $\mathrm{C} = \mathrm{A} - \mathrm{F}a$, so that the *definite integral*, as it is then called, is $\mathrm{F}x + \mathrm{A} - \mathrm{F}a$.

We shall now proceed to integrate a few fundamental expressions.

Integration of the form $(\mathrm{F}x)^n d\mathrm{F}x.$

(4.) This differential obviously corresponds to the differential of $(\mathrm{F}x)^{n+1} + \mathrm{C}$, with the exception that it is not multiplied by $n + 1$. If, therefore, we multiply it by this factor, and then place the reciprocal of it outside the sign $\int$, the expression under the sign will be thus rendered integrable,

$$\therefore \frac{1}{n+1} \int (n+1)\,(Fx)^n\,dFx = \frac{(Fx)^{n+1}}{n+1} + C.$$

Hence, to integrate any differential of the proposed form, that is, where the expression without the parenthesis is the differential of that within, the rule is to *increase the power of the function within the parenthesis by unity, divide this increased power by its exponent, and annex the arbitrary constant.* We shall subjoin a few examples of expressions coming under this form, or which may be easily reduced to it.

EXAMPLES.

(5.) 1. To integrate ax^5dx.

$$\int ax^5\,dx = \frac{ax^6}{6} + C.$$

2. To integrate $\sqrt{a + x^2}\,x\,dx$ or $(a + x^2)^{\frac{1}{2}}\,x\,dx$.

Here the expression without the parenthesis is not the complete differential of that within, requiring to be multiplied by 2; hence, introducing this factor, and placing its reciprocal outside, we have

$$\tfrac{1}{2}\int (a + x^2)^{\frac{1}{2}}\,2x\,dx = \frac{(a + x^2)^{\frac{3}{2}}}{\frac{3}{2}} + C.$$

3. To integrate $(b + cx^n)^m\,ax^{n-1}\,dx$.

Here it is easy to perceive that the expression without the parenthesis requires to be multiplied by $\dfrac{nc}{a}$; hence

$$\frac{a}{nc}\int (b + cx^n)^m\,ncx^{n-1}\,dx = \frac{a}{nc\,(m+1)}\,(b + cx^n)^{m+1} + C.$$

4. To integrate

$$\frac{adx}{(b - cx)^n} \text{ or } (b - cx)^{-n}\,adx,$$

$$-\frac{a}{c}\int (b - cx)^{-n}\,cdx = \frac{a\,(b - cx)^{1-n}}{c\,(n-1)} + C.$$

5. To integrate

$$dy = (ax + bx^2 + cx^3)^m\,(a + 2bx + 3cx^2)\,dx.$$

This being of the proposed form the integral is

$$y = \frac{(ax + bx^2 + cx^3)^{m+1}}{m+1} + C.$$

6. To integrate

$$dy = (2ax - x^2)^{\frac{5}{2}} (a - x)\, dx$$

$$y = \tfrac{1}{2} \int (2ax - x^2)^{\frac{5}{2}} (2a - 2x)\, dx = \frac{(2ax - x^2)^{\frac{7}{2}}}{7} + C.$$

In the examples already given it has been an easy matter to discover the factor necessary to render the expression without the parentheses the differential of that within; but there is a general method of ascertaining whether a proposed differential belongs to the case we are considering which ought to be noticed. Thus, taking the last example, assume

$$y = A\, (2ax - x^2)^{\frac{7}{2}} + C,$$

then, differentiating each member,

$$dy = (2ax - x^2)^{\frac{5}{2}} (a - x)\, dx = 7\, A\, (2ax - x^2)^{\frac{5}{2}} (2a - 2x)\, dx,$$

consequently, if the differential is of the proposed form, we must have the conditions

$$a = 7Aa, \; 1 = 7A,$$

which agree in giving the same value to A, viz. $A = \tfrac{1}{7}$; hence the integral is

$$y = \tfrac{1}{7} (2ax - x^2)^{\frac{7}{2}} + C,$$

as before determined.

If the example had been

$$dy = (2ax - x^2)^{\frac{3}{2}} (5a - x)\, dx,$$

then, as before, assuming

$$y = A\, (2ax - x^2)^{\frac{5}{2}},$$

and differentiating each member, we have

$$(2ax - x^2)^{\frac{3}{2}} (5a - x) = \tfrac{5}{2} A\, (2ax - x^2)^{\frac{3}{2}} (2a - 2x),$$

$$\therefore 5 = 5A, \; 1 = 5A,$$

two conditions which are contradictory; hence we infer that the differential does not belong to the proposed form.

7. To integrate

$$dy = adx - \frac{bdx}{x^3} + x^{\frac{3}{2}} dx$$

$$y = \int a\,dx - \int \frac{b\,dx}{x^3} + \int x^{\frac{3}{2}}\,dx = ax - \frac{b}{x^2} + \tfrac{2}{3} x^{\frac{5}{2}} + \mathrm{C}.$$

8. To integrate
$$dy = (a + bx)^2\,dx,$$
$$y = \int (a + bx)^2\,dx = \int (a^2\,dx + 2abx\,dx + b^2 x^2\,dx)$$
$$= a^2 x + abx^2 + \frac{b^2 x^3}{3} + \mathrm{C}.$$

9. To integrate
$$dy = \frac{2a\,dx}{x\,\sqrt{2ax - x^2}}.$$

This is the same as
$$x^{-1}(2ax - x^2)^{-\frac{1}{2}}\,2a\,dx = x^{-2}(2ax^{-1} - 1)^{-\frac{1}{2}}\,2a\,dx,$$

which is of the required form, with the exception of its sign,
$$\therefore y = -\int (2ax^{-1} - 1)^{-\frac{1}{2}} \times - 2ax^{-2}\,dx = - 2(2ax^{-1} - 1)^{\frac{1}{2}} + \mathrm{C}$$
$$= - 2\,\frac{\sqrt{2ax - x^2}}{x} + \mathrm{C}.$$

10. To integrate
$$dy = \frac{x\,dx}{(2ax - x^2)^{\frac{3}{2}}}$$

This is the same as
$$(2a - x)^{-\frac{3}{2}} x^{-\frac{1}{2}}\,dx = (2ax^{-1} - 1)^{-\frac{3}{2}} x^{-2}\,dx,$$

which will be of the required form, when the expression without the parenthesis is multiplied by $- 2a$; hence
$$y = - \frac{1}{2a}\int (2ax^{-1} - 1)^{-\frac{3}{2}} \times - 2ax^{-2}\,dx = \frac{(2ax^{-1} - 1)^{-\frac{1}{2}}}{a}$$
$$= \frac{1}{a}\sqrt{\frac{x}{2a - x}}.$$

11. To integrate
$$\frac{a\,dx}{x^3}, \quad \int \frac{a\,dx}{x^3} = \frac{3x^{\frac{5}{3}}}{5} + \mathrm{C}.$$

12. To integrate
$$dy = (a + bx + cx^2)^{\frac{2}{3}} (b\,dx + 2cx\,dx),$$
$$\therefore y = \tfrac{3}{5}(a + bx + cx^2)^{\frac{5}{3}} + \mathrm{C}.$$

13. To integrate

$$dy = (a + bx^2)^{\frac{1}{2}} \, mx dx.$$

$$\therefore y = \frac{m}{3b}(a + bx^2)^{\frac{3}{2}} + C.$$

14. To integrate

$$dy = 6\sqrt{4x^2 + 3} \cdot x dx,$$

$$\therefore y = \frac{1}{2}(4x^2 + 3)^{\frac{3}{2}} + C.$$

15. To integrate

$$dy = (a^2 + x^2)^2 \, x dx,$$

$$\therefore y = \frac{1}{6}(a^2 + x^2)^3.$$

16. To integrate

$$dy = \frac{x^8 dx}{\sqrt{a^9 + 6x^9}}$$

$$\therefore y = \frac{\sqrt{a^9 + 6x^9}}{27}.$$

17. To integrate

$$dy = \frac{6x^3(x^6 - 1)\, dx}{1 - x},$$

$$\therefore y = -\left(\frac{3x^4}{4} + \frac{6x^5}{5} + x^6 + \frac{6x^7}{7} + \frac{3x^8}{4} + \frac{2x^9}{3}\right).$$

18. To integrate

$$dy = \frac{\sqrt{a^2 + x^2}\, dx}{x^4},$$

$$\therefore y = -\frac{(a^2 + x^2)^{\frac{3}{2}}}{3a^2 x^3}.$$

(θ.) There is one case belonging to the above form, which never. theless does not correspond to the differential of any power, and to which, therefore, the foregoing rule does not apply. The case is that in which n becomes -1, the form being $\dfrac{dX}{X}$, which evidently agrees with the form for the differential of log. X, hence

$$\int \frac{dX}{X} = \log. X + C = \log. cX,$$

c being the number whose logarithm is C. The following examples belong to this case.

19. To integrate

$$\frac{3x^2dx}{x^3 + a^2}.$$

The numerator being the differential of the denominator, we have

$$\int \frac{3x^2dx}{x^3 + a^2} = \log. \, C \, (x^3 + a^2).$$

20. To integrate

$$\frac{adx}{a + bx}.$$

To render the numerator equal to the differential of the denominator, we must multiply it by $\dfrac{b}{a}$,

$$\therefore \frac{a}{b} \int \frac{bdx}{a + bx} = \frac{a}{b} \log. \, C \, (a + bx).$$

21. To integrate

$$\frac{ax^n dx}{b^n + cx^{n+1}}.$$

$$\frac{a}{(n+1)c} \int \frac{(n+1)\,cx^n dx}{b^n + cx^{n+1}} = \frac{a}{(n+1)c} \log. \, C \, (b^n + cx^{n+1}).$$

22. To integrate

$$\frac{5x^3 dx}{3x^4 + 7}.$$

$$\int \frac{5x^3 dx}{3x^4 + 7} = \frac{5}{12} \log. \, C \, (3x^4 + 7).$$

23. To integrate

$$\frac{\frac{1}{3}xdx}{x^2 + \frac{3}{4}}.$$

$$\int \frac{\frac{1}{3}xdx}{x^2 + \frac{3}{4}} = \frac{1}{6} \log. \, C \, (x^2 + \frac{3}{4}).$$

24. To integrate

$$dy = \frac{a}{b^4} \cdot \frac{(x - a)^4 dx}{x^2}.$$

$$\therefore y = \frac{a}{b^4} \left\{ \frac{x^2}{2} - 3ax + 3a^2 \log. \, x + \frac{a^3}{x} \right\} + C.$$

(7.) *Integration of the Forms*

$$\pm \int \frac{dx}{\sqrt{a^2 - b^2 x^2}} , \quad \pm \int \frac{dx}{a^2 + b^2 x^2} , \quad \pm \int \frac{dx}{x \sqrt{b^2 x^2 - a^2}} ,$$

$$\pm \int \frac{dx}{\sqrt{a^2 x - b^2 x^2}} .$$

If we put $\dfrac{a}{b} = r$, the three first of these forms will be the same as

$$\pm \frac{1}{a} \int \frac{r dx}{\sqrt{r^2 - x^2}}, \quad \pm \frac{1}{a^2} \int \frac{r^2 dx}{r^2 + x^2}, \quad \pm \frac{b}{a^2} \int \frac{r^2 dx}{x \sqrt{x^2 - r^2}},$$

where the expressions under the sign of integration are identical with those at article (16) in the *Differential Calculus.*

As to the fourth form, if we put $\dfrac{a^2}{b^2} = 2r$ it will be the same as

$$\pm \frac{2b}{a^2} \int \frac{r dx}{\sqrt{2rx - x^2}},$$

the expression under the sign of integration being identical to the remaining expression in the article just referred to. Hence the integrals of all the proposed forms are given by the circular arcs exhibited in that article, so that

$$\int \frac{dx}{\sqrt{a^2 - b^2 x^2}} = \frac{1}{a} \sin.^{-1} x + C$$

$$-\int \frac{dx}{\sqrt{a^2 - b^2 x^2}} = \frac{1}{a} \cos.^{-1} x + C$$

$$\int \frac{dx}{a^2 + b^2 x^2} = \frac{1}{a^2} \tan.^{-1} x + C$$

$$-\int \frac{dx}{a^2 + b^2 x^2} = \frac{1}{a^2} \cot.^{-1} x + C$$

$$\int \frac{dx}{x \sqrt{b^2 x^2 - a^2}} = \frac{b}{a^2} \sec.^{-1} x + C$$

$$\int \frac{dx}{x \sqrt{b^2 x^2 - a^2}} = \frac{b}{a^2} \operatorname{cosec.}^{-1} x + C$$

$$\int \frac{dx}{\sqrt{a^2 x - b^2 x^2}} = 2 \frac{b}{a^2} \operatorname{versin.}^{-1} x + C$$

$$-\int \frac{dx}{\sqrt{a^2x - b^2x^2}} = 2\,\frac{b}{a^2}\,\text{coversin.}^{-1}\,x + C,$$

where it must be observed that in all these expressions, except the last two, the radius is $\dfrac{a}{b}$ and in the last two it is $\dfrac{a}{2b}\sqrt{2}$.

(8.) If $a = b = 1$, then $r = 1$, and the first six integrals are, simply,

$$\int \frac{dx}{\sqrt{1 - x^2}} = \sin.^{-1}\,x + C$$

$$-\int \frac{dx}{\sqrt{1 - x^2}} = \cos.^{-1}\,x + C$$

$$\int \frac{dx}{1 + x^2} = \tan.^{-1}\,x + C$$

$$-\int \frac{dx}{1 + x^2} = \cot.^{-1}\,x + C$$

$$\int \frac{dx}{x\sqrt{x^2 - 1}} = \sec.^{-1}\,x + C$$

$$-\int \frac{dx}{x\sqrt{x^2 - 1}} = \text{cosec.}^{-1}\,x + C,$$

and if $b = 1$, $a^2 = 2$, then $r = 1$, and the last two are

$$\int \frac{dx}{\sqrt{2x - x^2}} = \text{versin.}^{-1}\,x + C$$

$$-\int \frac{dx}{\sqrt{2x - x^2}} = \text{coversin.}^{-1}\,x + C,$$

the radius of these arcs being all unity.

(9.) The more general expressions in art. (7) may also be so modified as to involve only the common tabular trigonometrical quantities, or those to radius 1. For, if any trigonometrical line belonging to an arc of radius r, be divided by r, the quotient will be the trigonometrical line belonging to a similar arc of radius 1, we have, therefore, merely to multiply this arc by r, to arrive at the arc of radius r originally proposed. Hence, if, in the expressions art. (7), we divide x by the radius to which it belongs and multiply the corresponding arc by that radius, the values of those expressions will remain unaltered, and will be calculable, for particular values of x, by means of the common trigonometrical tables. The expressions thus modified are

$$\int \frac{dx}{\sqrt{a^2 - b^2 x^2}} = \frac{1}{b} \sin.^{-1} \frac{b}{a} x + C$$

$$-\int \frac{dx}{\sqrt{a^2 - b^2 x^2}} = \frac{1}{b} \cos.^{-1} \frac{b}{a} x + + C$$

$$\int \frac{dx}{a^2 + b^2 x^2} = \frac{1}{ab} \tan.^{-1} \frac{b}{a} x + C$$

$$-\int \frac{dx}{a^2 + b^2 x^2} = \frac{1}{ab} \cot.^{-1} \frac{b}{a} x + C$$

$$\int \frac{dx}{x \sqrt{b^2 x^2 - a^2}} = \frac{1}{a} \sec.^{-1} \frac{b}{a} x + C$$

$$-\int \frac{dx}{x \sqrt{b^2 x^2 - a^2}} = \frac{1}{a} \operatorname{cosec}.^{-1} \frac{b}{a} x + C$$

$$\int \frac{dx}{\sqrt{a^2 x - b^2 x^2}} = \frac{1}{b} \operatorname{versin}.^{-1} \frac{2b^2}{a^2} x + C$$

$$-\int \frac{dx}{\sqrt{a^2 x - b^2 x^2}} = \frac{1}{b} \operatorname{coversin}.^{-1} x \frac{2b^2}{a^2} x + C.$$

(10.) These circular forms will repeatedly occur hereafter, and the student should endeavour to carry them in his mind. It is obvious that these same forms hold, if instead of x there be substituted any function of it X, as we shall now illustrate by a few examples.

EXAMPLES.

1. To integrate

$$\frac{x dx}{\sqrt{a - b x^4}}.$$

If, in this expresssion, X be put for x^2, we have $x dx = \frac{1}{2} dX$, therefore

$$\frac{1}{2} \int \frac{dX}{\sqrt{a - b X^2}} = \frac{1}{2 b^{\frac{1}{2}}} \sin.^{-1} \frac{b^{\frac{1}{2}}}{a^{\frac{1}{2}}} X + C$$

$$= \frac{1}{2b^{\frac{1}{2}}} \sin.^{-1} \frac{b^{\frac{1}{2}}}{a^{\frac{1}{2}}} x^2 + C.$$

2. To integrate

$$\frac{x^{n-1} dx}{\sqrt{a - bx^{2n}}}.$$

Putting $x^n = X$, we have $x^{n-1} dx = \frac{1}{n} dX$,

$$\therefore \frac{1}{n} \int \frac{dX}{\sqrt{a - bX^2}} = \frac{1}{nb^{\frac{1}{2}}} \sin.^{-1} \frac{b^{\frac{1}{2}}}{a^{\frac{1}{2}}} x^n.$$

3. To integrate

$$\frac{x^{\frac{1}{2}} dx}{a + bx^3}.$$

Putting $x^{\frac{3}{2}} = X$, we have $x^{\frac{1}{2}} dx = \frac{2}{3} dX$,

$$\therefore \frac{2}{3} \int \frac{dX}{a + bX^2} = \frac{2}{3\sqrt{ab}} \tan.^{-1} \sqrt{\frac{b}{a}} \cdot x^3.$$

4. To integrate

$$\frac{dx}{x\sqrt{bx^n - a}}.$$

Multiplying numerator and denominator by $x^{\frac{n}{2} - 1}$ this expression becomes

$$\frac{x^{\frac{n}{2} - 1}}{x^{\frac{n}{2}} \sqrt{bx^n - a}},$$

and, putting $x^{\frac{n}{2}} = X$, we have $x^{\frac{n}{2} - 1} = \frac{2}{n} dX$,

$$\therefore \frac{2}{n} \int \frac{dX}{X \sqrt{bX^2 - a}} = \frac{2}{na^{\frac{1}{2}}} \sec.^{-1} \sqrt{\frac{b}{a}} x^n.$$

5. To integrate

$$\frac{dx}{\sqrt{ax + bx^2}}.$$

Dividing numerator and denominator by $x^{\frac{1}{2}}$, the express,ou becomes

$$\frac{x^{-\frac{1}{2}}\,dx}{\sqrt{a+bx}}.$$

and, putting $x^{\frac{1}{2}} = \mathrm{X}$, we have $x^{-\frac{1}{2}} = -2\,d\mathrm{X}$,

$$\therefore -2 \int \frac{d\mathrm{X}}{\sqrt{a+b\mathrm{X}^2}} = \frac{2}{b^{\frac{1}{2}}} \cos.^{-1} \sqrt{\frac{b}{a}}\, x.$$

6. To integrate

$$\frac{x\,dx}{1+x^4},$$

$$\int \frac{x\,dx}{1+x^4} = \tfrac{1}{2} \tan.^{-1} x^2.$$

7. To integrate

$$\frac{x^{\frac{1}{2}}\,dx}{\sqrt{2-4x^3}},$$

$$\int \frac{x^{\frac{1}{2}}\,dx}{\sqrt{2-4x^3}} = \tfrac{1}{3} \sin.^{-1} \sqrt{2x^3}.$$

8. To integrate

$$\frac{x^{n-1}\,dx}{1+x^{2n}}.$$

$$\int \frac{x^{n-1}\,dx}{1+x^{2n}} = \frac{1}{n} \tan.^{-1} x^n.$$

(11.) We have now, by inverting a few of the fundamental processes of the differential calculus, shewn how to integrate the most simple forms of those differential expressions which lead to algebraical, logarithmic, and circular functions. It remains to consider those which depend upon exponential, and trigonometrical functions; still, however, confining ourselves to the most simple forms of those expressions that can possibly occur; we shall thus have all the elementary forms of which the most complicated integral can be composed. The exponential and trigonometrical forms are as follow:

Since $da^x = \log.a \,.\, a^x\,dx$ therefore $\int a^x\,dx = \dfrac{a^x}{\log.a} + \mathrm{C}$

$de^x = e^x dx \quad . . . . \int e^x\,dx = e^x + \mathrm{C}$

$d\sin.x = \cos.x\,dx \quad . . . . \int \cos.x\,dx = \sin.x + \mathrm{C}$

$$d \cos x = -\sin x \, dx \quad \ldots \ldots \quad \int \sin x \, dx = -\cos x + C$$

$$d \tan x = \frac{dx}{\cos.^2 x} \quad \ldots \ldots \ldots \quad \int \frac{dx}{\cos.^2 x} = \tan x + C$$

$$d \cot x = \frac{dx}{\sin.^2 x} \quad \ldots \ldots \ldots \quad \int \frac{dx}{\sin.^2 x} = \cot x + C$$

$$d \sec x = \tan x \sec x \, dx \quad \ldots \ldots \quad \int \tan x \sec x \, dx = \sec x + C$$

$$d \csc x = -\cot x \csc x \, dx \quad \ldots \int \cot x \csc x \, dx = -\csc x + C.$$

Having thus collected together in the present chapter all the elementary forms, our principal object throughout this section will now be to decompose into these forms every differential whose integral we wish to determine.

CHAPTER II.

ON THE INTEGRATION OF RATIONAL FRACTIONS.

(10.) By the aid of the elementary integrals, determined in last chapter, we may integrate every differential contained in the general form

$$\frac{P x^{m-1} + Q x^{m-3} \ldots \ldots + R x + S}{P' x^m + Q' x^{m-1} \ldots \ldots \ldots + R'} \, dx,$$

provided we can by any means decompose the denominator of the fractional coefficient into its simple or quadratic factors.

In the form here exhibited we see that the highest exponent of x in the numerator is less than the highest exponent of x in the denominator by at least one unit, but if any rational fraction be proposed having the highest exponent of x in the numerator greater than the highest exponent in the denominator, then, by actually performing the division indicated, we shall obtain a quotient of the form $p x^n dx$, and a remainder, in which the highest exponent of x is less than the highest exponent in the divisor; the fraction, therefore, formed by this remainder and divisor will be of the above form, and this fraction annexed to the quotient must be equal to the proposed; we shall have, therefore, to integrate these two parts, and as the first belongs to the form (4) there will remain to be integrated the form above, so that the integration of this form comprehends the integration of every form of the rational fraction.

(11.) To shew in the simplest manner how this integration is to be effected we shall apply the process to particular examples, choosing at first those fractions of which the factors of the denominator are all unequal.

 1. Let it be required to integrate

$$\frac{a}{x^2 - a^2}\, dx.$$

The factors of the denominator are here $x - a$ and $x + a$, and our object is now to find what two partial fractions $\dfrac{A}{x - a}$ and $\dfrac{B}{x + a}$ compose the proposed, that is to say, what values of A and B satisfy the condition

$$\frac{a}{x^2 - a^2} = \frac{A}{x - a} + \frac{B}{x + a}.$$

By reducing the two partial fractions to a common denominator, and actually adding the numerators, this condition reduces to

$$a = (A + B)\, x + (A - B)\, a,$$

and as this must exist, whatever be the value of x, we have, by the method of indeterminate coefficients,

$$A + B = 0,\ a = (A - B)\, a \therefore 1 = A - B,$$

which equations give

$$A = \tfrac{1}{2},\ B = -\tfrac{1}{2};$$

hence the partial fractions are determined, and we have

$$\int \frac{a}{x^2 - a^2}\, dx = \tfrac{1}{2} \int \frac{dx}{x - a} - \tfrac{1}{2} \int \frac{dx}{x + a},$$

that is (6),

$$\int \frac{a\, dx}{x^2 - a^2} = \tfrac{1}{2} \log. (x - a) - \tfrac{1}{2} \log. (x + a) + C$$

$$= \log. \ (\frac{x-a}{x+a})^{\frac{1}{2}} + C.$$

2. Let it be required to integrate

$$\frac{a^3 + bx^2}{a^2x - x^3} \, dx.$$

In this example the factors of the denominator are x, $a - x$, and $a + x$, and in order to decompose the fractional coefficient into partial fractions, we must so determine A, B, and C, that we may have the condition

$$\frac{a^3 + bx^2}{a^2x - x^3} = \frac{A}{x} + \frac{B}{a - x} + \frac{C}{a + x},$$

which, as in last example, reduces to the condition

$$a^3 + bx^2 = Aa^2 - Ax^2 + Bax + Bx^2 + Cax - Cx^2,$$

therefore, equating the coefficients of the like powers of x, we have the equations

$$B - A - C = b, \ Ba + Ca = 0, \ Aa^2 = a^3.$$

The last of these immediately gives $A = a$, which reduces the first to $B - C = a + b$; also, since the second is the same as $B + C = 0$, we get for B and C the values

$$B = \frac{a+b}{2}, \ C = - \frac{a+b}{2};$$

the partial fractions being thus determined, we have

$$\int \frac{a^3 + bx^2}{a^2x - x^3} dx = a \int \frac{dx}{x} + \frac{a+b}{2} \int \frac{dx}{a-x} - \frac{a+b}{2} \int \frac{dx}{a+x}$$

$$= a \log. x - \frac{a+b}{2} \log. (a - x) - \frac{a+b}{2} \log. (a + x) + C$$

$$= a \log. x - \frac{a+b}{2} \log. (a^2 - x^2) + C.$$

3. Let it be required to integrate

$$\frac{3x - 5}{x^2 - 6x + 8} \, dx.$$

To determine the factors of the denominator we must find the roots of the equation

$$x^2 - 6x + 8 = 0,$$

which are $x = 2$ and $x = 4$; hence the factors are $x - 2$ and $x - 4$; therefore, as before, assuming

$$\frac{3x - 5}{x^2 - 6x + 8} = \frac{A}{x - 2} + \frac{B}{x - 4},$$

we have the condition

$$3x - 5 = Ax - 4A + Bx - 2B,$$

therefore, comparing the like powers of x, we have the equations

$$3 = A + B, \; 5 = 4A + 2B,$$

$$\therefore A = -\tfrac{1}{2}, \; B = \tfrac{7}{2},$$

consequently

$$\int \frac{3x - 5}{x^2 - 6x + 8}\, dx = -\tfrac{1}{2} \int \frac{dx}{x - 2} + \tfrac{7}{2} \int \frac{dx}{x - 4}$$

$$= \tfrac{7}{2} \log. (x - 4) - \tfrac{1}{2} \log. (x - 2) + C.$$

4. Let it be required to integrate

$$\frac{x}{x^2 + 4ax - b^2}\, dx.$$

Decomposing the denominator, as in the last example, we find for the factors

$$x + 2a + \sqrt{4a^2 + b^2} \text{ and } x + 2a - \sqrt{4a^2 + b^2}.$$

or, more briefly,

$$x + K \text{ and } x + L;$$

hence, assuming

$$\frac{x}{x^2 + 4ax - b^2} = \frac{A}{x + K} + \frac{B}{x + L};$$

we have the condition

$$x = Ax + AL + B + BK,$$

which furnishes the equations

$$A + B = 1, \; AL + BK = 0,$$

from which we find for A and B the values

$$A = \frac{K}{K - L},\ B = -\frac{L}{K - L}$$

$$\therefore \int \frac{x}{x^2 + 4ax - b^2}\,dx = \frac{K}{K - L}\int \frac{dx}{x + K} - \frac{L}{K - L}\int \frac{dx}{x + L}$$

$$= \frac{K}{K - L}\log.(x + K) - \frac{L}{K - L}\log.(x + L) + C.$$

5. To integrate

$$\frac{a}{x^2 - 5x + 6}\,dx$$

$$\int \frac{a}{x^2 - 5x + 6}\,dx = d\log.\frac{x - 3}{x - 2} + C.$$

6. To integrate

$$\frac{2x + 3}{x^3 + x^2 - 2x}\,dx$$

$$\int \frac{2x + 3}{x^3 + x^2 - 2x}\,dx = \tfrac{5}{3}\log.(x - 1) - \tfrac{1}{6}\log.(x + 2) - \tfrac{3}{2}\log. x.$$

7. To integrate

$$\frac{2 - 4x}{x^2 - x - 2}\,dx$$

$$\int \frac{2 - 4x}{x^2 - x - 2}\,dx = -2\log.(x^2 - x - 2).$$

From these examples it appears that when the denominator of the rational fraction can be decomposed into *simple* and *unequal* factors, the integral of the expression will always be determinable, and will always be of a *logarithmic form*, because the several component partial differentials will be fractions whose numerators are the differentials of the denominators, whether these be rational or imaginary.

(12.) When the factors of the denominator are not only simple and rational, but some of them equal, the process just employed must be modified a little. Thus, suppose we had to decompose the fraction

$$\frac{a + bx + cx^2}{(x - k)^3}$$

where the factors of the denominator are all equal.

The partial fractions cannot here be of the form

$$\frac{A}{x-k}, \quad \frac{B}{x-k}, \quad \frac{C}{x-k}$$

for, as these are all of the same denominator, their sum is of the form

$$\frac{A'}{x-k}$$

and the condition for determining the numerators

$$a + bx + cx^2 = A'(x-k)^2,$$

which is not only insufficient for that purpose, but it also fixes a relation between k and a, b, c,.

If, however, we make, in the proposed expression, this substitution, viz,

$$x - k = z \therefore x = z + k$$

it will take the form

$$\frac{a + bk + ck^2 + bz + 2ckz + cz^2}{z^3}$$

of which the component fractions are obviously

$$\frac{a + bk + ck^2}{z^3}, \quad \frac{b + 2ck}{z^2}, \quad \frac{c}{z}.$$

Hence the component fractions of the proposed are

$$\frac{x + bk + ck^2}{(x-k)^3}, \quad \frac{b + 2ck}{(x-k)^2}, \quad \frac{c}{x-k},$$

that is

$$\frac{a + bx + cx^2}{(x-k)^3} = \frac{A}{(x-k)^3} + \frac{B}{(x-k)^2} + \frac{C}{x-k}.$$

It is easy to perceive, from the process employed in this instance, that a similar form of decomposition has place in every case where the denominator of the rational fraction consists of only equal rational and simple factors. When unequal factors enter as well, the corresponding partial fractions will be determined, as in the case already considered; but the operations will here, as in that case, be best understood by means of a few particular examples:

8. Let it be required to integrate

$$\frac{2ax}{(x+a)^2}dx.$$

Here we have to determine A and B from the condition

$$\frac{2ax}{(x+a)^2} = \frac{A}{(x+a)^2} + \frac{B}{x+a},$$

which, by actually adding the fractions in the second member, and equating the numerators, leads to

$$2ax = A + Bx + Ba,$$

which gives the equations

$$2a = B, \ A + Ba = 0,$$

that is,

$$A = -2a^2, \ B = 2a,$$

$$\therefore \int \frac{2ax}{(x+a)^2}dx = -2a^2 \int \frac{dx}{(x+a)^2} + 2a \int \frac{xdx}{x+a}$$

$$= -\frac{2a^2}{x+a} + 2a \log.(x+a) + C.$$

9. Let it be required to integrate

$$\frac{x^2}{x^3 - ax^2 - a^2x + a^3}dx.$$

In order to decompose the denominator, we must find the roots of the equation

$$x^3 - ax^2 - a^2x + a^3 = 0;$$

it is easy, however, to see that $x = a$ is one of these roots; therefore, depressing the equation by the easy method explained at page 193 of my Algebra, there results the quadratic factor $x^2 - a^2$, which gives the simple factors $x - a$, $x + a$; hence, assuming

$$\frac{x^2}{(x-a)^2(x+a)} = \frac{A}{(x-a)^2} + \frac{B}{x-a} + \frac{C}{x+a};$$

reducing the partial fractions to a common denominator, and equating the numerators, we have

$$1 = B + C, \ A - 2Ca = 0, \ Aa - Ba^2 + Ca^2 = 0.$$

If we multiply the first of these conditions by a^2, and add the result to the third, we shall have

$$Aa + 2Ca^2 = a^2,$$

and adding this to the second, multiplied by a, there results

$$a^2 = 2Aa \therefore A = \tfrac{1}{2}\, a,$$

this, substituted in the second condition, gives

$$C = \tfrac{1}{4},$$

whence the first reduces to

$$B = 1 - \tfrac{1}{4} = \tfrac{3}{4},$$

therefore

$$\int \frac{x^2}{(x-a)^2(x+a)}\, dx = \frac{a}{2} \int \frac{dx}{(x-a)^2} + \frac{3}{4} \int \frac{dx}{x-a}\; \frac{1}{4} \int \frac{dx}{x+a}$$

$$= -\frac{a}{2(x-a)} + \frac{3}{4} \log.(x-a) + \frac{1}{4} \log.(x+a) + C.$$

10. Let it be required to integrate

$$\frac{a}{(x^2-1)^2}\, dx = \frac{a}{(x-1)^2(x+1)^2}\, dx.$$

Here we must assume

$$\frac{a}{(x-1)^2(x+1)^2} = \frac{A}{(x-1)^2} + \frac{B}{x-1} + \frac{C}{(x+1)^2} + \frac{D}{x+1};$$

and, by reducing the partial fractions to a common denominator, we are led to these equations of condition, viz.

$$B + D = 0$$

$$A + B + C - D = 0$$

$$2A - B - 2C - D = 0$$

$$A - B + C + D = a.$$

The first of these reduces the third to $2A - 2C = 0$, therefore $A = C$, the second reduces the fourth to $2A + 2C = a$, therefore, since $A = C$, $A = \tfrac{1}{4} a = C$, consequently the fourth becomes $D - B = \tfrac{1}{2}a$, which, combined with the first, gives

$$B = -\tfrac{1}{4}\, a,\; D = \tfrac{1}{4}\, a;$$

hence

$$\int \frac{a\,dx}{(x^2-1)^2} = \tfrac{1}{4}\,a \int \left\{ \frac{dx}{(x+1)^2} + \frac{dx}{(x+1)^2} - \frac{dx}{x-1} + \frac{dx}{x+1} \right\}$$

$$= \tfrac{1}{4}\,a \left\{ -\frac{1}{x-1} - \frac{1}{x+1} - \log.(x-1) + \log.(x+1) \right\} + C.$$

It appears from these examples that when the denominator of the rational fraction has all its simple factors rational and some of them equal, the integral is determinable, and can consist only of algebraic and logarithmic functions. The result is the same if imaginary factors enter, but we prefer to make this a distinct case.

Before examining the case in which the denominator of the fractional coefficient contains imaginary factors, we shall add a few more examples, in the two cases already considered, for the exercise of the student.

11. To integrate

$$\frac{2a}{a^2-x^2}\,dx.$$

$$\int \frac{2a}{a^2-x^2}\,dx = \log.\frac{a+x}{a-x} + C.$$

12. To integrate

$$\frac{dx}{x^3 - 7x^2 + 12x}.$$

$$\int \frac{dx}{x^3 - 7x^2 + 12x} = \tfrac{1}{12}\log. x + \tfrac{1}{4}\log.(x-4) - \tfrac{1}{3}\log.(x-3) + C.$$

13. To integrate

$$\frac{x^2 dx}{x^3 - x^2 - x + 1};$$

$$\int \frac{x^2 dx}{x^3 - x^2 - x + 1} = \frac{1}{2-2x} + \tfrac{3}{4}\log.(x-1) + \tfrac{1}{4}\log.(x+1) + C.$$

14. To integrate

$$\frac{x\,dx}{(a+bx)^3}.$$

$$\int \frac{x\,dx}{(a+bx)^3} = -\frac{a+2bx}{2b^2(a+bx)^2}.$$

15. To integrate

$$\frac{x^2 - 2}{x^3 + 4x^2 + 4x} \, dx.$$

$$\int \frac{x^2 - 2}{x^3 + 4x^2 + 4x} \, dx = \frac{1}{x + 2} + \log. \frac{(x+2)^{\frac{3}{2}}}{x^{\frac{1}{4}}} + C.$$

16. To integrate

$$\frac{x^2 - 3}{x^3 - 7x + 6} \, dx.$$

$$\int \frac{x^2 - 3}{x^3 - 7x + 6} \, dx = \tfrac{1}{2} \log. (x - 1) + \tfrac{1}{3} \log. (x - 2) + \tfrac{3}{10} \log. (x + 3) + C.$$

(13.) It remains to consider the case in which imaginary factors enter the denominator.

The imaginary roots of an equation always occur in pairs and are of the forms

$$x = a + \beta \sqrt{-1} \text{ and } x = a - \beta \sqrt{-1};$$

so that the quadratic factor which gives these roots is of the form

$$x^2 - 2ax + a^2 + \beta^2 = (x - a)^2 + \beta^2,$$

and, therefore, the corresponding partial fraction of the form

$$\frac{Mx + N}{(x - a)^2 + \beta^2},$$

which cannot be decomposed into rational partial fractions; but, if there enter into the denominator of the proposed several equal quadratic factors of this kind, or, which is the same thing, if there enters as a factor the power

$$\{(x - a)^2 + \beta^2\}^m,$$

the corresponding partial fraction will be of the form

$$\frac{Px^{2m-1} + Qx^{2m-2} \ldots\ldots\ldots + W}{\{(x - a)^2 + \beta^2\}^m} \ \ldots\ldots (1),$$

or, by introducing the indeterminate coefficients A, B, C, &c. these may be so determined as to render this fraction identical to

$$\frac{Ax + B + (Cx + D) \{(x - a)^2 + \beta^2\} + (Ex + F) \{(x - a)^2 + \beta^2\}^2 + \&c.}{\{(x - a)^2 + \beta^2\}^m}$$

the last factor in the numerator being

$$(Ix + K)\,\{(x - a)^2 + \beta^2\}^{m-1};$$

hence the partial fraction (1) is equal to the sum of the fractions

$$\frac{Ax + B}{\{(x - a)^2 + \beta^2\}^m} + \frac{Cx + D}{\{(x - a)^2 + \beta^2\}^{m-1}} + \cdots \frac{Ix + K}{(x - a)^2 + \beta^2} \cdots (2).$$

Knowing, therefore, the form of the component partial fractions, we may readily analyse any rational fraction when we can find the simple factors of its denominator, whether these be rational or imaginary.

From the form of decomposition just established when equal quadratic factors enter the denominator, it is obviously necessary, in order to complete the integration of the class of differentials considered in this chapter, without using imaginaries, that we know how to integrate the form

$$\frac{Ax + B}{\{(x - a)^2 + \beta^2\}^m}\,dx,$$

which, by putting z for $x - a$, becomes

$$\frac{Az + Aa + B}{(z^2 + \beta^2)^m}\,dz,$$

or substituting a for $Aa + B$,

$$\frac{Az\,dz}{(z^2 + \beta^2)^m} + \frac{a\,dz}{(z^2 + \beta^2)^m}.$$

The first of these forms we know how to integrate, having considered it in (4), its integral is

$$\frac{A}{2}\int \frac{2z\,dz}{(z^2 + \beta^2)^m} = -\frac{A}{2\,(m - 1)\,(z^2 + \beta^2)^{m-1}},$$

it remains, therefore, to integrate the form

$$\frac{a\,dz}{(z^2 + \beta^2)^m} \cdots \cdots (3.)$$

Now this integration we cannot immediately effect, but it is easy to shew that the integral could be obtained, provided we could integrate

$$\frac{dz}{(z^2 + \beta^2)^{m-1}} \cdots \cdots (4),$$

because, if we multiply both numerator and denominator of this by $x^2 + \beta^2$, we have

$$\frac{dz}{(z^2+\beta^2)^{m-1}} = \frac{z^2dz}{(z^2+\beta^2)^m} + \frac{\beta^2dz}{(z^2+\beta^2)^m} \cdot \cdot \cdot \cdot (5),$$

and if to this equation we add

$$d \cdot \frac{z}{(z^2+\beta^2)^{m-1}} = \frac{dz}{(z^2+\beta^2)^{m-1}} - \frac{2(m-1)z^2dz}{(z^2+\beta^2)^m},$$

the integrals of the results are

$$\frac{z}{(z^2+\beta^2)^{m-1}} + \int \frac{dz}{(z^2+\beta^2)^{m-1}} = \int \frac{dz}{(z^2+\beta^2)^{m-1}} -$$

$$(2m-3)\int \frac{z^2dz}{(z^2+\beta^2)^m} + \beta^2 \int \frac{dz}{(z^2+\beta^2)^m}$$

$$\therefore \int \frac{z^2dz}{(z^2+\beta^2)^m} = -\frac{z}{(2m-3)(z^2+\beta^2(^{m-1}} + \frac{\beta^2}{2m-3}\int \frac{dz}{(z^2+\beta^2)^m};$$

substituting this value in the integral of (5), there results

$$\int \frac{dz}{(z^2+\beta^2)^{m-1}} = -\frac{z}{(2m-3)(z^2+\beta^2)^{m-1}} +$$

$$\frac{(2m-2)\beta^2}{2m-3}\int \frac{dz}{(z^2+\beta^2)^m},$$

and consequently

$$\int \frac{dz}{(z^2+\beta^2)^m} = \frac{z}{\beta^2(2m-2)(z^2+\beta^2)^{m-1}} +$$

$$\frac{2m-3}{\beta^2(2m-2)}\int \frac{dz}{(z^2+\beta^2)^{m-1}} \cdot \cdot \cdot \cdot (6).$$

Hence, as remarked above, the integral of (3) depends on the integral of (4), and, by the same formula, if $m-1$ be substituted for m, the integral of

$$\frac{dz}{(z^2+\beta^2)^{m-1}}$$

will become dependent on that of

$$\frac{dz}{(z^2+\beta^2)^{m-2}}$$

so that, by first determining the integral of $\dfrac{dz}{z^2+\beta^2}$, which we already know to be (7), $\dfrac{1}{\beta} \tan.^{-1} \dfrac{z}{\beta}$ we may, by the formula (6), determine in succession

$$\int \frac{dz}{(z^2 + \beta^2)^2} , \int \frac{dz}{(z^2 + \beta^2)^3} , \ldots \ldots \int \frac{dz}{(z^2 + \beta^2)^m} .$$

We shall, in the next chapter, shew how such integrals may be obtained, by another and more general process; we see here that they always involve a circular function. The following are examples of these integrals.

(14.) 17. Let it be required to integrate

$$\frac{x}{x^3 - 1} dx.$$

The factors of the denominator being

$$x - 1 \text{ and } x^2 + x + 1,$$

the latter involving imaginary factors of the first degree, the form of the decomposition is

$$\frac{x \, dx}{x^3 - 1} = \frac{A dx}{x - 1} + \frac{Bx + C}{x^2 + x + 1} dx;$$

and from this equation we are to determine A, B, C; therefore, reducing to a common denominator, and equating the like powers of x, in the numerators, as in the former examples, we have the conditions,

$$A + B = 0, \ A + C - B = 1, \ A - C = 0.$$

If we add these three equations together, we get

$$A = \tfrac{1}{3} \therefore B = -\tfrac{1}{3}, \ C = -\tfrac{1}{3},$$

consequently

$$\int \frac{x}{x^3 - 1} dx = \tfrac{1}{3} \int \frac{dx}{x - 1} - \tfrac{1}{3} \int \frac{x - 1}{x^2 + x + 1} dx.$$

The first of these component integrals is $\tfrac{1}{3}$ log. $(x - 1)$ and the second, being put under the form

$$\tfrac{1}{3} \int \frac{x - 1}{(x + \tfrac{1}{2})^2 + \tfrac{3}{4}} dx,$$

and z being substituted for $x + \tfrac{1}{2}$, becomes,

$$-\frac{1}{3}\int\frac{z\,dz}{z^2+\frac{3}{4}}+\frac{1}{2}\int\frac{dz}{z^2+\frac{3}{4}}=$$

$$-\frac{1}{6}\log.\,(z^2+\frac{3}{4})+\frac{1}{\sqrt{3}}\tan.^{-1}\frac{2}{\sqrt{3}}\,z+C\,;$$

hence, restoring the value of z, and collecting together the three component integrals, we have

$$\int\frac{x}{x^3-1}\,dx=\tfrac{1}{3}\{\log.\,(x-1)-\tfrac{1}{2}\log.\,(x^2+x+1)+$$

$$\sqrt{3}\tan.^{-1}\frac{2x+1}{\sqrt{3}}\}+C.$$

18. To integrate

$$\frac{x^4+2x^3+3x^2+3}{(x^2+1)^3}\,dx.$$

Here we must assume

$$\frac{x^4+2x^3+3x^2+3}{(x^2+1)^3}=\frac{Ax+B}{(x^2+1)^3}+\frac{Cx+D}{(x^2+1)^2}+\frac{E}{x^2+1},$$

from which we get, by actually adding the partial fractions and equating the numerators,

$$x^4+2x^3+3x^2+3=$$

$$Ex^4+Cx^3+(D+2E)\,x^2+(A+C)\,x+B+D+E,$$

consequently

$$E=1,\,C=2\,\therefore\,D=1,\,A=-2,\,B=1\,;$$

hence we have to determine

$$-2\int\frac{x\,dx}{(x^2+1)^3}+2\int\frac{x\,dx}{(x^2+1)^2}+\int\frac{dx}{(x^2+1)^3}+\int\frac{dx}{(x^2+1)^2}$$

$$+\int\frac{dx}{x^2+1}.$$

The two first of these integrals are, omitting the arbitrary constants,

$$\frac{1}{2\,(x^2+1)^2}\,\text{ and }\,-\frac{1}{x^2+1}\,;$$

also, by the formula (6),

$$\int \frac{dx}{(x^2+1)^3} = \frac{x}{4\,(x^2+1)^2} + \frac{3}{4}\int \frac{dx}{(x^2+1)^2}$$

$$\frac{7}{4}\int \frac{dx}{(x^2+1)^2} = \frac{7x}{8\,(x^2+1)} + \frac{7}{8}\int \frac{dx}{x^2+1}$$

$$\frac{15}{8}\int \frac{dx}{x^2+1} = \frac{15}{8}\tan.^{-1} x.$$

The first row of vertical terms on the right hand of the signs of equality are together equal to the sum of the three remaining integrals, therefore adding these to the two already determined, we have

$$\int \frac{x^4+2x^3+3x^2+3}{(x^2+1)^3}\,dx = \frac{2+x}{4\,(x^2+1)^2} + \frac{7x-8}{8\,(x^2+1)} +$$

$$\frac{15}{8}\tan.^{-1} x + C.$$

19. To integrate

$$\frac{x^2-x+1}{x^3+x^2+x+1}\,dx.$$

$$\int \frac{x^2-x+1}{x^3+x^2+x+1}\,dx = \frac{3}{2}\log.(x+1) - \frac{1}{4}\log.(x^2+1) -$$

$$\frac{1}{2}\tan.^{-1} x + C.$$

20. To integrate

$$\frac{a+bx}{x^3-1}\,dx.$$

$$\int \frac{a+bx}{x^3-1}\,dx = \frac{a+b}{3}\log.\frac{x-1}{\sqrt{x^2+x+1}} + \frac{b-a}{\sqrt{3}}\tan.^{-1}\frac{x+\frac{1}{2}}{\sqrt{\frac{3}{4}}} + C.$$

From what has now been done it appears that the integral of any differential whose coefficient is a rational fraction can always be determined by means of the elementary *algebraic, logarithmic,* and *tangential* forms, provided we can decompose the denominator of the fractional coefficient into its constituent factors. There are several irrational forms which may be rationalized by means of certain transformations and reductions, and which may, therefore, be integrated by aid of the principles already laid down. We shall now consider the principal of these irrational forms to which general processes apply.

Reduction of Irrational Functions to Rational.

(15.) The simplest irrational function which can occur is that which consists of monomial terms only, and these are very easily rationalized; it will be necessary merely to reduce the fractional indices of the variable x to a common denominator m, and then to substitute z^m for x.

1. Suppose, for example, the differential

$$\frac{x^{\frac{1}{2}} - \frac{1}{3}a}{x^{\frac{1}{3}} - x^{\frac{1}{2}}}\, dx$$

were proposed, then, since the common denominator given by the reduction of the fractions $\frac{1}{2}$, $\frac{1}{3}$, is 6, we must substitute z^6 for x, which substitution reduces the differential to the rational form

$$\frac{z^3 - \frac{1}{3}a}{z^2 - z^3}\, 6z^5 dz = \frac{6z^6 - 2az^3}{1 - z}\, dz.$$

Since the highest exponent of z in the numerator of this expression exceeds the highest exponent in the denominator, we must perform the actual division, by which we get

$$\frac{6z^6 - 2az^3}{1 - z}\, dz = \left\{ -6z^5 - 6z^4 - 6z^3 - (6-2a)z^2 - (6-2a)z - (6-2a) \right\} dz$$
$$+ \frac{6 - 2a}{1 - z}\, dz,$$

consequently

$$\int \frac{6z^6 - 2az^3}{1 - z}\, dz = - z^6 - \frac{6}{5}z^5 - \frac{3}{2}z^4 - \frac{6 - 2a}{3}z^3 - (3-a)z^2 -$$
$$(6 - 2a)z + (2a - 6) \log. (z - 1) + C.$$

In this manner it is obvious that we may render rational and then integrate every differential included in the general form

$$\frac{ax^{\frac{m}{n}} + bx^{\frac{p}{q}} + \&c.}{a'x^{\frac{m'}{n'}} + b'x^{\frac{p'}{q'}} + \&c.}\, dx.$$

2. As a second example, the student may take the differential

$$\frac{x^{\frac{1}{2}} - 2x^{\frac{1}{3}}}{1 + x^{\frac{1}{3}}} dx,$$

the integral of which will be found to be

$$\int \frac{x^{\frac{1}{2}} - 2x^{\frac{1}{3}}}{1 + x^{\frac{1}{3}}} = \frac{6x^{\frac{7}{6}}}{7} - 2x - \frac{6x^{\frac{5}{6}}}{5} + 3x^{\frac{2}{3}} + 2x^{\frac{1}{2}} - 6x^{\frac{1}{3}} - 6x^{\frac{1}{6}} +$$

$$\log. (x^{\frac{1}{3}} + 1) + \tan.^{-1} x^{\frac{1}{6}}.$$

(16.) If the surds which enter the function, instead of being mono-mial, are binomial, and all of the form $(a + bx)^{\frac{m}{n}}$, the function may likewise be rationalized. For if, as before, we reduce all the fractional exponents of $(a + bx)$ to a common denominator p, and then assume $a + bx = z^p$, the coefficient of dx will obviously be rational, and dx will become $\dfrac{pz^{p-1} dz}{b}$, which is also rational; hence such a function will be thus rendered entirely rational.

1. Suppose for example the differential proposed were

$$\frac{dx}{\sqrt{a + bx}}.$$

Putting $a + bx = z^2$ we have

$$dx = \frac{2z dz}{b},$$

$$\therefore \frac{dx}{\sqrt{a + bx}} = \frac{2 dz}{b};$$

$$\therefore \int \frac{dx}{\sqrt{a + bx}} = 2 \int \frac{dz}{b} = 2 \frac{z}{b} + C = \frac{2 \sqrt{a + bx}}{b} + C.$$

2. Again, let it be required to integrate

$$\frac{x dx}{(1 + x)^{\frac{3}{2}}}.$$

Substituting z^2 for $1 + x$, we have $x = z^2 - 1 \therefore x dx = 2 (z^2 - 1) z dz$; hence

$$\frac{x\,dx}{(1+x)^{\frac{3}{2}}} = \frac{2\,(z^2-1)\,dz}{z^2} = 2dz - \frac{2dz}{z^2}$$

$$\therefore \int \frac{x\,dx}{(1+x)^{\frac{3}{2}}} = 2z + \frac{2}{z} + C = 2\left\{\sqrt{1+x} + \frac{1}{\sqrt{1+x}}\right\} + C.$$

3. As a third example let it be proposed to integrate

$$\frac{dx}{x\sqrt{a+bx}}.$$

Here the transformed expression in z will be found to be

$$\frac{2dz}{z^2-a} = \frac{2dz}{z^2+(\sqrt{-a})^2},$$

the integral of which is either

$$\frac{2}{\sqrt{a}}\,\log.\,\frac{\sqrt{a+bx}-\sqrt{a}}{x}\ \text{or}\ \frac{2}{\sqrt{-a}}\,\tan.^{-1}\sqrt{\frac{a+bx}{-a}},$$

according as the first or second form is taken. If the differential had been

$$\frac{dx}{x\sqrt{bx-a}},$$

then the expression in z would have been

$$\frac{2dz}{z^2+a}\ \text{or}\ \frac{2dz}{z^2-(\sqrt{-a})^2},$$

of which the integral is

$$\frac{2}{\sqrt{a}}\,\tan.^{-1}\sqrt{\frac{bx-a}{a}}\ \text{or}\ \frac{2}{\sqrt{-a}}\,\log.\,\frac{\sqrt{bx-a}-\sqrt{-a}}{x}.$$

It is thus obvious that we may always give to the integral of

$$\frac{bdz}{z^2 \pm a}$$

either a logarithmic or a circular form, whichever we please, but one of these forms will necessarily involve imaginaries, and will therefore be in general less suitable for the purposes of calculation than the other.

(17.) There remains one more class of irrational differentials which can be rendered rational by a general process; these are such as involve no other irrational terms but those of the form

$$\sqrt{a + bx + cx^2},$$

which form may in every case be rendered rational, by applying the principles of the diophantine analysis, and consequently every rational function of it may be rendered rational.

First, let c be positive, and assume

$$\frac{a}{c} + \frac{bx}{c} + x^2 = (x + z)^2 = x^2 + 2xz + z^2,$$

from which equation we get

$$x = \frac{a - cz^2}{2cz - b} \quad \therefore \quad dx = -\frac{2c(a - bz + cz^2)}{(2zc - b)^2} dz;$$

hence

$$\sqrt{a + bx + cx^2} = (x + z)\sqrt{c} = \frac{a - bz + cz^2}{2cz - b}\sqrt{c};$$

consequently, by the proposed transformation we obtain for the irrational function of x an equivalent rational function of z, and as also x itself is a rational function of z, dx must be a rational function of z; so that differentials of the proposed form are thus rendered entirely rational.

Secondly, let c be negative, and let α and β be the two roots of the equation

$$x^2 - \frac{bx}{c} - \frac{a}{c} = 0,$$

then, by changing the signs

$$\frac{a}{c} + \frac{bx}{c} - x^2 = -(x - \alpha)(x - \beta) = (x - \alpha)(\beta - x),$$

having thus decomposed the expression under the radical into its simple factors, we shall assume

$$\sqrt{(x - \alpha)(\beta - x)} = (x - \alpha)z,$$

from which we get

$$\beta - x = (x - \alpha)z^2,$$

whence

$$x = \frac{\beta + \alpha z^2}{z^2 + 1} \quad \therefore \quad dx = \frac{2(\alpha - \beta)z}{(z^2 + 1)^2} dz;$$

therefore

$$x - a = \frac{\beta + az^2}{z^2 + 1} - a = \frac{\beta - a}{z^2 + 1};$$

hence, by substitution in the original assumption, we have

$$\sqrt{(x - a)(\beta - x)} = \frac{\beta - a}{z^2 + 1} z;$$

which is a rational expression, and so likewise is the expression for x, and therefore the expression for dx must be rational too.

It appears that when, in such irrational differentials, as we are now considering, the coefficient of x^2 is negative, it will be necessary, in order to rationalize them, to determine the roots of the quadratic function under the radical, after changing the signs of the terms; but when the coefficient of x^2 is positive, this preliminary operation will be unnecessary. We shall add an example or two of these forms:

(18.)   4. Let it be proposed to integrate

$$\frac{dx}{\sqrt{a + cx^2}}.$$

Here we have to rationalize

$$\sqrt{a + 0x + cx^2};$$

therefore, proceeding as in the first case above, we have

$$x = \frac{a - cz^2}{2cz}, dx = -\frac{a + cz^2}{2cz^2}dz, \sqrt{a + cx^2} = \frac{a + cz^2}{2\sqrt{c}z}:$$

and, consequently,

$$\int \frac{dx}{\sqrt{a + cx^2}} = \int \frac{-dz}{\sqrt{c}z} = \frac{-1}{\sqrt{c}} \log. z,$$

that is, since

$$z = \frac{1}{\sqrt{c}}\sqrt{a + cx^2} - x$$

$$\int \frac{dx}{\sqrt{a + cx^2}} = -\frac{1}{\sqrt{c}} \log. C \{\sqrt{a + cx^2} - x\sqrt{c}\}.$$

As the sum of the squares of the terms within the brackets is $= a$, if we divide by the constant log. a, which we may incorporate with the arbitrary constant log. C, the form will be changed into

$$\int \frac{dx}{\sqrt{a+cx^2}} = \frac{1}{\sqrt{c}} \log. \ C \ \{\sqrt{a+cx^2} + x\sqrt{c}\},$$

and a similar change may be effected on the integral in the next example.

If we put the proposed differential under the form

$$\frac{dx}{a-(\sqrt{-c})^2 x^2};$$

the corresponding form of the integral will be

$$\frac{1}{\sqrt{-c}} \sin.^{-1} \frac{\sqrt{-cx}}{a} + C.$$

5. To integrate

$$\frac{dx}{\sqrt{a+bx+cx^2}}.$$

Proceeding as above, we have

$$\int \frac{dx}{\sqrt{a+bx+cx^2}} = -\int \frac{2c^{\frac{1}{2}}dz}{2cz-b} = -\frac{1}{\sqrt{c}} \log. \ C \ (2cz-b),$$

that is, substituting for z its value

$$z = \frac{\sqrt{a+bx+cx^2}}{\sqrt{c}} - x$$

$$\int \frac{dx}{\sqrt{a+bx+cx^2}} = -\frac{1}{\sqrt{c}} \log. \ C \ \{2\sqrt{c(a+bx+cx^2)} - 2cx - b\}.$$

6. To integrate

$$\frac{dx}{\sqrt{a+bx-x^2}}.$$

Having determined the roots a, β of the equation

$$x^2 - bx - a = 0,$$

we have, by proceeding agreeably to the second case, above,

$$dx = -\frac{2(\beta-a)z}{(z^2+1)^2}, \quad \sqrt{a+bx-x^2} = \frac{\beta-a}{z^2+1} z;$$

consequently

$$\int \frac{dx}{\sqrt{a+bx-x^2}} = -\int \frac{2dz}{z^2+1} = -2\tan.^{-1} z + C,$$

or, restoring the value of z,

$$\int \frac{dx}{a + bx - x^2} = C - 2 \tan.^{-1} \sqrt{\frac{\beta - x}{x - a}}.$$

7. To determine

$$\int \frac{dx}{\sqrt{a^2 + b^2 x^2}}.$$

$$\int \frac{dx}{\sqrt{a^2 + b^2 x^2}} = \frac{1}{b} \log. C \left(bx + \sqrt{a^2 + b^2 x^2} \right)$$

8. To determine

$$\int \frac{dx}{x \sqrt{a^2 + b^2 x^2}}.$$

$$\int \frac{dx}{x \sqrt{a^2 + b^2 x^2}} = - \frac{1}{a} \log. C \frac{\sqrt{a^2 + b^2 x^2} + a}{x}$$

9. To integrate

$$\frac{dx}{(1 + x^2) \sqrt{1 - x^2}}.$$

$$\int \frac{dx}{(1 + x^2) \sqrt{1 - x^2}} = \frac{1}{\sqrt{2}} \tan.^{-1} \frac{x \sqrt{2}}{\sqrt{1 - x^2}} + C.$$

10. To integrate

$$\frac{dx}{x \sqrt{1 + x + x^2}}.$$

$$\int \frac{dx}{x \sqrt{1 + x + x^2}} = - \log. \frac{2 + x + 2 \sqrt{1 + x + x^2}}{x} C.$$

11. To integrate

$$\frac{dx}{\sqrt{2bx - x^2}}.$$

$$\int \frac{dx}{\sqrt{2bx - x^2}} = \cos.^{-1} \frac{b - x}{b} + C.$$

12. To integrate

$$\frac{adx}{\sqrt{2ax + x^2}}.$$

$$\int \frac{adx}{\sqrt{2ax + x^2}} = a \log. C \left(x + a + \sqrt{2ax + x^2} \right),$$

Besides the irrational forms considered in this chapter, there are others also reducible, by general rules, to rational forms. These, however, being all only particular cases of a more general form to be examined in the next chapter, they more properly come under notice in that place. We ought, perhaps, before dismissing the subject of this chapter to apprise the student that there exists another method of determining the coefficients A, B, C, &c. in the numerators of the assumed partial fractions, which does not require the equations of condition necessary in the method of indeterminate coefficients which we have employed. But, although this second method is in some cases shorter than that we have adopted, yet, as it is less simple and obvious, we have preferred the latter. The other method is explained in note (A), at the end of the volume, to which the student may refer.

CHAPTER III.

ON THE INTEGRATION OF BINOMIAL DIFFERENTIALS IN GENERAL.

(19.) The object of the present chapter is to solve the following general problem, viz.

To integrate the form

$$x^m (a + bx^n)^{\frac{p}{q}} \, dx \; \ldots \ldots \; (\mathrm{A}),$$

in which m, n, p, are either whole or fractional, positive or negative.

We shall first remark that this general expression may always be changed into another in which p shall be the only fractional exponent, and in which n shall be positive. For if we reduce the exponents m, n to a common denominator q, and then substitute z^q for x, p will be the only fractional exponent in the transformed expression: if after this the exponent of z within the parenthesis should be negative, we have only to substitute $\dfrac{1}{y}$ for z and it will become positive; hence the integration of every differential of the above form may be obtained provided we can always integrate when n is integral and positive, m being either a positive or a negative integer, and p any number whatever; so that, in fact, we need consider the above form only under these conditions, although, in what follows, this is not necessary.

Substitute z for $a + bx^n$ and there results

$$x = \left(\frac{z - a}{b}\right)^{\frac{1}{n}}$$

$$\therefore x^m = \left(\frac{z - a}{b}\right)^{\frac{m}{n}} \quad \therefore x^{m-1}\, dx = \frac{1}{nb^{\frac{m}{n}}} (z - a)^{\frac{m}{n} - 1}\, dz$$

$$\therefore x^m\, dx = \frac{1}{nb^{\frac{m+1}{n}}} (z - a)^{\frac{m+1}{n} - 1}\, dz;$$

hence the general form becomes

$$\frac{1}{nb^{\frac{m+1}{n}}} (z - a)^{\frac{m+1}{n} - 1} z^p\, dz.$$

Now if $\dfrac{m + 1}{n}$ should happen to be a whole number or 0, the exponent of $(z - a)$ will be a whole number r, and we shall then merely have to integrate the form

$$(z - a)^r\, z^p\, dz,$$

which we can always do whether r is positive or negative; for if it is positive $(z - a)^r$ is, when developed, a series of monomials, and thus the integration is finally dependent on the form $z^s\, dz$; if r is negative and t be the denominator of the fraction p, then, by substituting y^t for z, the form is reduced to a rational fraction. Hence *the form may always be rendered rational when* $\dfrac{m + 1}{n}$ *is an integer.* This is called the condition of integrability.

(21.) By adopting a little artifice we may easily arrive at another transformation of the general differential expression, and thence obtain another condition of integrability. Thus, divide one of the factors $(a + bx^n)^p$ of the proposed by x^{np} and it becomes $(ax^{-n} + b)^p$; multiply the other factor x^m by x^{np} and it becomes x^{m+np}, so that the proposed is the same as

$$x^{m+np} (ax^{-n} + b)^p \, dx.$$

Substitute in this z for $ax^{-n} + b$ and the resulting transformed expression can differ from that before obtained only in this, that a and b will be interchanged, that $-n$ will appear instead of n, and $m + np$ instead of m; hence the transformed expression will here be

$$- \frac{a^{\frac{m+1}{n} + p}}{n} \cdot (z - b)^{-\frac{m+1}{n} - p - 1} z^p \, dz.$$

Hence *the form may be rendered rational when* $\dfrac{m + 1}{n} + p$ *is an integer or 0.*

The foregoing are the only cases of the general form which in the present state of analysis can be rendered rational. The following examples satisfy the conditions of integrability.

EXAMPLES.

(22.) 1. To determine the integral of

$$x^3 (a + bx^2)^{\frac{1}{2}} \, dx.$$

In this example

$$m = 3, \; n = 2, \; p = \tfrac{1}{2} \text{ and } \frac{m + 1}{n} = 2,$$

the first of the preceding conditions is therefore satisfied, and the transformed differential is

$$\frac{1}{2b^2} (z - a) z^{\frac{1}{2}} \, dx,$$

in which

$$z = a + bx^2;$$

consequently, taking the integral

E

$$\frac{z^{\frac{5}{2}}}{5b^2} - \frac{az^{\frac{3}{2}}}{3b^2} + C = \left(\frac{a+bx^2}{5} - \frac{a}{3}\right) \frac{\{a+bx^2\}^{\frac{3}{2}}}{b^2} + C.$$

2. It is required to integrate

$$x^{-2}\,(a+x^3)^{-\frac{5}{3}}\,dx$$

$$m = -2,\; n = 3,\; p = -\frac{5}{3},\quad \frac{m+1}{n} + p = -2;$$

hence the second condition of integrability is satisfied, and the transformed differential is

$$-\frac{a^{-2}}{3}\,(z-1)\,z^{-\frac{5}{3}}\,dz,$$

of which the integral is

$$-\frac{1}{a^2}\,(z^{\frac{1}{3}} + \tfrac{1}{2}\,z^{-\frac{2}{3}}),$$

where

$$z = ax^{-3} + 1 = \frac{a+x^3}{x^3},$$

consequently,

$$\int x^{-2}\,(a+x^3)^{-\frac{5}{3}}\,dx = \frac{\dfrac{a+x^3}{x^3} + \tfrac{1}{2}}{a^2\dfrac{(a+x^3)^{\frac{2}{3}}}{x^2}} + C$$

$$= C - \frac{3x^3 + 2a}{2a^2 x\,(a+x^3)^{\frac{2}{3}}}.$$

3. To integrate

$$x^3\,(a^2+x^2)^{\frac{1}{3}}\,dx$$

$$\int x^3\,(a^2+x^2)^{\frac{1}{3}}\,dx = \frac{3}{56}\,(a^2+x^2)^{\frac{4}{3}}\,(4x^2 - 3a^2) + C.$$

4. To integrate

$$\frac{a\,dx}{(1+x^2)^{\frac{3}{2}}}$$

$$\int \frac{a\,dx}{(1+x^2)^{\frac{3}{2}}} = \frac{ax}{\sqrt{1+x^2}} + C.$$

5. To integrate

$$x^5\,(a+bx^2)^{\frac{2}{3}}\,dx$$

$$\int x^5 (a + bx^2)^{\frac{2}{3}} \, dx = \frac{3z^{\frac{5}{3}}}{2b^3} \left(\frac{z^2}{11} - \frac{az}{4} + \frac{a^2}{5} \right) + C,$$

in which

$$z = a + bx^2.$$

(23.) When the conditions of integrability are not satisfied, the proposed differential may then be referred to other general formulas called *formulas of reduction*, and which reduce the integration of the proposed expression to others of a simpler kind. These formulas are obtained as follows: From the known form

$$duv = udv + vdu$$

we have, by taking the integrals

$$uv = \int udv + \int vdu$$

$$\therefore \int udv = uv - \int vdu \ \dots \dots (1),$$

a formula which reduces the integration of *udv* to that of **vdu**, and which is known by the name of *integration by parts*.

Let us now compare $\int udv$ with the integral $\int x^m (a + bx^n)^p \, dx$ in supposing

$$(a + bx^n)^p = u, \quad x^m dx = dv \ \therefore \ v = \frac{x^{m+1}}{m+1},$$

and we shall then have, by applying the method of integration by parts,

$$\int x^m (a + bx^n)^p \, dx =$$

$$\left(a + bx^n \right)^p \frac{x^{m+1}}{m+1} - \frac{pnb}{m+1} \int x^{m+1} (a + bx^n)^{p-1} x^{n-1} \, dx,$$

or putting as before

$$a + bx^n = z$$

$$\int x^m z^p \, dx = z^p . \frac{x^{m+1}}{m+1} - \frac{pnb}{m+1} \int x^{m+n} z^{p-1} \, dx \ \dots \dots (2).$$

By this formula of reduction we see that the integral of any differential of the form (A) is made to depend upon the integral of another differential of the same form, but in which the exponent of z is diminished by 1, and the exponent of x, without the parenthesis, increased by n.

From this we may deduce a second formula, for since

$$z^p = z^{p-1}(a + bx^n) = az^{p-1} + bz^{p-1}x^n$$

it follows that

$$\int x^m z^p \, dx = a\int x^m z^{p-1} \, dx + b\int x^{m+n} z^{p-1} \, dx \ . \ . \ . \ . \ (3).$$

Subtract this from equation (2) and there results

$$a\int x^m z^{p-1} \, dx = z^p \frac{x^{m+1}}{m+1} - \frac{(pn + m + 1)\,b}{m+1}\int x^{m+n} z^{p-1} \, dx,$$

or, substituting p for $p - 1$,

$$\int x^m z^p \, dx = z^{p+1} \frac{x^{m+1}}{a(m+1)} - \frac{(pn + n + m + 1)\,b}{a\,(m+1)}\int x^{m+n} z^p \, dx,$$

by which formula the integral is made to depend upon another of the same form, but in which the exponent of x, without the parenthesis, is increased by n.

The two formulas now given may obviously be useful when m is negative; it may be remarked, however, that both fail to be applicable when $m + 1 = 0$, or when $m = -1$, but in this case they are not wanted, because as then $\dfrac{m+1}{n} = 0$ the condition of integrability is satisfied, and the proposed form may therefore be rendered rational.

If we transpose the integrals in the formula last deduced we shall have

$$\int x^{m+n} z^p \, dx = z^{p+1} \cdot \frac{x^{m+1}}{(pn + n + m + 1)\,b}$$

$$- \frac{a\,(m+1)}{(pn + n + m + 1)b}\int x^m z^p \, dx,$$

which, by putting m instead of $m + n$, becomes

$$\int x^m z^p \, dx = z^{p+1} \cdot \frac{x^{m-n+1}}{(pn + m + 1)\,b} - \frac{a\,(m - n + 1)}{(pn + m + 1)\,b}\int x^{m-n} z^p \, dx$$

a formula which causes the proposed integral to depend on another of the same form, but having the exponent of x without the parenthesis diminished by n.

If instead of subtracting equation (3) from equation (2) we had multiplied it by $\dfrac{pn}{m+1}$ and then added, we should have had

$$\left(1 + \frac{pn}{m+1}\right) \int x^m z^p \, dx = z^p \cdot \frac{x^{m+1}}{m+1} + \frac{apn}{m+1} \int x^m z^{p-1} \, dx,$$

whence, dividing by the coefficient $\dfrac{pn + m + 1}{m+1}$, we have

$$\int x^m z^p \, dx = z^p \cdot \frac{x^{m+1}}{pn + m + 1} + \frac{apn}{pn + m + 1} \int x^m z^{p-1} \, dx,$$

by which formula the integral depends on another having the exponent of the binomial less by unity.

By multiplying this last formula by the denominator and transposing the integrals, we have

$$\int x^m z^{p-1} \, dx = - z^p \cdot \frac{x^{m+1}}{apn} + \frac{pn + m + 1}{apn} \int x^m z^p \, dx$$

which, by putting p for $p-1$, becomes

$$\int x^m z^p \, dx = - z^{p+1} \cdot \frac{x^{m+1}}{a(p+1)n} + \frac{(p+1)n + m + 1}{a(p+1)n} \int x^m z^{p+1} \, dx$$

a formula which may be useful when p is negative.

In like manner, by multiplying the formula (2) by $\dfrac{m+1}{pnb}$ and transposing the integrals, we have

$$\int x^{m+n} z^{p-1} \, dx = z^p \cdot \frac{x^{m+1}}{pnb} - \frac{m+1}{pnb} \int x^m z^p \, dx,$$

which, by putting m instead of $m+n$ and p instead of $p-1$, becomes

$$\int x^m z^p \, dx = z^{p+1} \cdot \frac{x^{m-n+1}}{(p+1)nb} - \frac{m-n+1}{(p+1)nb} \int x^{m-n} z^{p+1} \, dx.$$

For the convenience of reference we shall now collect together the several formulas deduced in this article, and we shall thus have the following

(24.) TABLE OF FORMULAS FOR THE REDUCTION OF THE INTEGRAL,

$$x^m (a + bx^n)^p \, dx \text{ or } \int x^m z^p \, dx.$$

I.

$$\int x^m z^p \, dx = z^p \cdot \frac{x^{m+1}}{pn + m + 1} + \frac{apn}{pn + m + 1} \int x^m z^{p-1} \, dx.$$

II.

$$\int x^m z^p \, dx = z^p \cdot \frac{x^{m+1}}{m + 1} - \frac{pnb}{m + 1} \int x^{m-n} z^{p+1} \, dx.$$

III.

$$\int x^m z^p \, dx = - z^{p+1} \cdot \frac{x^{m+1}}{a (p + 1) n} + \frac{(p + 1) n + m + 1}{a (p + 1) n} \int x^m z^{p+1} \, dx.$$

IV.

$$\int x^m z^p \, dx = z^{p+1} \cdot \frac{x^{m-n+1}}{(p + 1) nb} - \frac{m - n + 1}{(p + 1) nb} \int x^{m-n} z^{p+1} \, dx.$$

V.

$$\int x^m z^p \, dx = z^{p+1} \cdot \frac{x^{m-n+1}}{(pn + m + 1) b} - \frac{a (m - n + 1)}{(pn + m + 1) b} \int x^{m-n} z^p \, dx.$$

VI.

$$\int z^m z^p \, dx = z^{p+1} \cdot \frac{x^{m+1}}{a (m + 1)} - \frac{(pn + n + m + 1) b}{a (m + 1)} \int x^{m+n} z^p \, dx.$$

Either of these formulas may under certain relations of the exponents become inapplicable on account of the denominator vanishing, but it will be easy to perceive that under these same relations the differentials proposed may be rendered rational. We shall now apply the foregoing formulas of reduction to some examples.

EXAMPLES.

(25.) 1. To integrate

$$\frac{x^5\,dx}{x^2+a^2}=x^5\,(x^2+a^2)^{-1}\,dx.$$

To this expression we may conveniently apply the formula V., from which we have

$$\int\frac{x^5\,dx}{x^2+a^2}=\frac{x^4}{4}-a^2\int\frac{x^3\,dx}{x^2+a^2}$$

$$\int\frac{x^3\,dx}{x^2+a^2}=\frac{x^2}{2}-a^2\int\frac{x\,dx}{x^2+a^2},$$

we have thus reduced the integral to the known form

$$a^2\int\frac{x\,dx}{x^2+a^2}=\frac{a^2}{2}\,\log.\,(x^2+a^2),$$

therefore

$$\int\frac{x^5\,dx}{x^2+a^2}=\frac{x^4}{4}-\frac{a^2\,x^2}{2}+\frac{a^4}{2}\,\log.\,(x^2+a^2)+\mathrm{C}.$$

2. To integrate

$$\frac{x^5\,dx}{\sqrt{a+bx^2}}.$$

Applying to this expression the same formula we have

$$\int\frac{x^5\,dx}{\sqrt{a+bx^2}}=\int x^5\,z^{-\frac{1}{2}}\,dx=z^{\frac{1}{2}}\cdot\frac{x^4}{5b}-\frac{4a}{5b}\int x^3\,z^{-\frac{1}{2}}\,dx$$

$$\int x^3\,z^{-\frac{1}{2}}\,dx=z^{\frac{1}{2}}\cdot\frac{x^2}{3b}-\frac{2a}{3b}\int x\,z^{-\frac{1}{2}}\,dx,$$

also

$$\int x\,(a+bx^2)^{-\frac{1}{2}}\,dx=\frac{\sqrt{a+bx}}{b},$$

consequently

$$\int\frac{x^5\,dx}{\sqrt{a+bx^2}}=\frac{x^4\sqrt{a+bx^2}}{5b}-\frac{4a}{15b^2}x^2\sqrt{a+bx^2}+\frac{8a^2}{15b^3}\sqrt{a+bx^2}+\mathrm{C}$$

$$=\frac{\sqrt{a+bx^2}}{15b}\left\{3x^4-\frac{4ax^2}{b}+\frac{8a^2}{b^2}\right\}+\mathrm{C}.$$

3. To integrate

$$\frac{x^m\,dx}{\sqrt{1-x^2}}.$$

By the same formula (V.) we have

$$\int\frac{x^m\,dx}{\sqrt{1-x^2}}=-\frac{1}{m}\,x^{m-1}\sqrt{1-x^2}+\frac{m-1}{m}\int\frac{x^{m-2}\,dx}{\sqrt{1-x^2}}\cdots (1),$$

and making m successively equal to the odd numbers 1, 3, &c. this equation gives

$$\int\frac{x\,dx}{\sqrt{1-x^2}}=-\sqrt{1-x^2}+C$$

$$\int\frac{x^3\,dx}{\sqrt{1-x^2}}=-\frac{1}{3}\,x^2\sqrt{1-x^2}+\frac{2}{3}\int\frac{x\,dx}{\sqrt{1-x^2}}$$

$$\int\frac{x^5\,dx}{\sqrt{1-x^2}}=-\frac{1}{5}\,x^4\sqrt{1-x^2}+\frac{4}{5}\int\frac{x^3\,dx}{\sqrt{1-x^2}}$$

$$\int\frac{x^7\,dx}{\sqrt{1-x^2}}=-\frac{1}{7}\,x^6\sqrt{1-x^2}+\frac{6}{7}\int\frac{x^5\,dx}{\sqrt{1-x^2}}$$

$$\vdots$$

$$\int\frac{x^m\,dx}{\sqrt{1-x^2}}=-\frac{1}{m}\,x^{m-1}\sqrt{1-x^2}+\frac{m-1}{m}\int\frac{x^{m-2}\,dx}{\sqrt{1-x^2}}.$$

Substituting in each of the right hand members the value of the integral as given by the preceding equation we have

$$\int\frac{x\,dx}{\sqrt{1-x^2}}=-\sqrt{1-x^2}+C$$

$$\int\frac{x^3\,dx}{\sqrt{1-x^2}}=-\left(\frac{1}{3}\,x^2+\frac{1\cdot 2}{1\cdot 3}\right)\sqrt{1-x^2}+C$$

$$\int\frac{x^5\,dx}{\sqrt{1-x^2}}=-\left(\frac{1}{5}\,x^4+\frac{1\cdot 4}{3\cdot 5}\,x^2+\frac{1\cdot 2\cdot 4}{1\cdot 3\cdot 5}\right)\sqrt{1-x^2}+C$$

$$\int\frac{x^7\,dx}{\sqrt{1-x^2}}=-\left(\frac{1}{7}\,x^6+\frac{1\cdot 6}{5\cdot 7}\,x^4+\frac{1\cdot 4\cdot 6}{3\cdot 5\cdot 7}\,x^2+\right.$$
$$\left.\frac{1\cdot 2\cdot 4\cdot 6}{1\cdot 3\cdot 5\cdot 7}\right)\sqrt{1-x^2}+C$$

$$\int \frac{x^m\,dx}{\sqrt{1-x^2}} = C - \left\{ \frac{1}{m} x^{m-1} + \frac{m-1}{(m-2)m} x^{m-3} + \right.$$

$$\frac{(m-3)(m-1)}{(m-4)(m-2)m} x^{m-5} + \ldots$$

$$\left. \ldots\ldots\ldots + \frac{1\cdot2\cdot4\cdot6\ldots\ldots(m-1)}{1\cdot3\cdot5\cdot7\ldots\ldots\ m} \right\} \sqrt{1-x^2}.$$

Let now m be assumed successively equal to the even numbers 0, 2, 4, &c. For $m=0$ the formula is inapplicable, but the integral for this case is given at art. (8), and is $\sin.^{-1} x + C$; therefore

$$\int \frac{dx}{\sqrt{1-x^2}} = \sin.^{-1} x + C$$

$$\int \frac{x^2\,dx}{\sqrt{1-x^2}} = -\frac{1}{2} x \sqrt{1-x^2} + \frac{1}{2} \int \frac{dx}{\sqrt{1-x^2}}$$

$$\int \frac{x^4\,dx}{\sqrt{1-x^2}} = -\frac{1}{4} x^3 \sqrt{1-x^2} + \frac{3}{4} \int \frac{x^2\,dx}{\sqrt{1-x^2}}$$

$$\int \frac{x^6\,dx}{\sqrt{1-x^2}} = -\frac{1}{6} x^5 \sqrt{1-x^2} + \frac{5}{6} \int \frac{x^4\,dx}{\sqrt{1-x^2}}$$

$$\int \frac{x^m\,dx}{\sqrt{1-x^2}} = -\frac{1}{m} x^{m-1} \sqrt{1-x^2} + \frac{m-1}{m} \int \frac{x^{m-2}\,dx}{\sqrt{1-x^2}},$$

that is, substituting for the integral in each right hand member its value given by the preceding equation

$$\int \frac{dx}{\sqrt{1-x^2}} = \sin.^{-1} x + C$$

$$\int \frac{x^2\,dx}{\sqrt{1-x^2}} = -\frac{1}{2} x \sqrt{1-x^2} + \frac{1}{2} \sin.^{-1} x + C$$

$$\int \frac{x^4\,dx}{\sqrt{1-x^2}} = -\left(\frac{1}{4} x^3 + \frac{1\cdot3}{2\cdot4} x\right) \sqrt{1-x^2} + \frac{1\cdot3}{2\cdot4} \sin.^{-1} x + C$$

$$\int \frac{x^6\,dx}{\sqrt{1-x^2}} = -\left(\frac{1}{6} x^5 + \frac{1\cdot5}{4\cdot6} x^3 + \frac{1\cdot3\cdot5}{2\cdot4\cdot6} x\right) \sqrt{1-x^2} +$$

$$\frac{1\cdot 3\cdot 5}{2\cdot 4\cdot 6}\sin.^{-1}x + C$$

$$\int\frac{x^m\,dx}{\sqrt{1-x^2}} = C - \left\{\frac{1}{m}x^{m-1} + \frac{m-1}{(m-2)m}x^{m-3} + \right.$$

$$\frac{(m-3)(m-1)}{(m-4)(m-2)m}x^{m-5} +$$

$$\cdots\cdots\cdots + \frac{1\cdot 3\cdot 5\,\ldots\ldots\,(m-1)}{2\cdot 4\cdot 6\,\ldots\ldots\ldots\,m}\left.\right\}\sqrt{1-x^2} +$$

$$\frac{1\cdot 3\cdot 5\,\ldots\ldots\,(m-1)}{2\cdot 4\cdot 6\,\ldots\ldots\,m}\sin.^{-1}x + C$$

4. To integrate

$$\frac{dx}{x^m\sqrt{1-x^2}},$$

By the formula VI. we have

$$\int\frac{dx}{x^m\sqrt{1-x^2}} = -\frac{\sqrt{1-x^2}}{(m-1)x^{m-1}} + \frac{m-2}{m-1}\int\frac{dx}{x^{m-2}\sqrt{1-x^2}}\cdot\ (2),$$

which for $m = 1$ fails to be applicable; but example (8), art. (18),

$$\int\frac{dx}{x\sqrt{1-x^2}} = -\log.\frac{1+\sqrt{1-x^2}}{x} + C;$$

hence, putting m successively equal to 1, 2, 3, &c. we have

$$\int\frac{dx}{x\sqrt{1-x^2}} = -\log.\frac{1+\sqrt{1-x^2}}{x} + C$$

$$\int\frac{dx}{x^3\sqrt{1-x^2}} = -\frac{\sqrt{1-x^2}}{2x^2} + \frac{1}{2}\int\frac{dx}{x\sqrt{1-x^2}}$$

$$\int\frac{dx}{x^5\sqrt{1-x^2}} = -\frac{\sqrt{1-x^2}}{4x^4} + \frac{3}{4}\int\frac{dx}{x^3\sqrt{1-x^2}}$$

$$\int\frac{dx}{x^7\sqrt{1-x^2}} = -\frac{\sqrt{1-x^2}}{6x^6} + \frac{5}{6}\int\frac{dx}{x^5\sqrt{1-x^2}}$$

$$\vdots$$

$$\int \frac{dx}{x^m \sqrt{1-x^2}} = - \frac{\sqrt{1-x^2}}{(m-1)\,x^{m-1}} + \frac{m-2}{m-1} \int \frac{dx}{x^{m-2}\sqrt{1-x^2}}$$

that is, by substitution,

$$\int \frac{dx}{x\,\sqrt{1-x^2}} = - \log. \frac{1+\sqrt{1-x^2}}{x} + C$$

$$\int \frac{dx}{x^3\,\sqrt{1-x^2}} = - \frac{\sqrt{1-x^2}}{2x^2} - \frac{1}{2}\log. \frac{1+\sqrt{1-x^2}}{x} + C.$$

$$\int \frac{dx}{x^5\,\sqrt{1-x^2}} = - \left(\frac{1}{4x^4} + \frac{1\cdot 3}{2\cdot 4x^2}\right) \sqrt{1-x^2}$$
$$\qquad - \frac{1\cdot 3}{2\cdot 4}\log. \frac{1+\sqrt{1-x^2}}{x} + C$$

$$\int \frac{dx}{x^7\,\sqrt{1-x^2}} = - \left(\frac{1}{6x^6} + \frac{1\cdot 5}{4\cdot 6x^4} + \frac{1\cdot 3\cdot 5}{2\cdot 4\cdot 6x^2}\right) \sqrt{1-x^2}$$
$$\qquad - \frac{1\cdot 3\cdot 5}{2\cdot 4\cdot 6}\log. \frac{1+\sqrt{1-x^2}}{x} + C$$

$$\vdots$$

$$\int \frac{dx}{x^m \sqrt{1-x^2}} = C - \left\{ \frac{1}{(m-1)\,x^{m-1}} + \frac{m-2}{(m-3)(m-1)\,x^{m-3}} \right.$$
$$\qquad + \frac{(m-4)(m-2)}{(m-5)(m-3)(m-1)x^{m-5}} +$$
$$\qquad \cdots\cdots + \left. \frac{1\cdot 3\cdot 5\cdots\cdots(m-2)}{2\cdot 4\cdot 6\cdots(m-1)\,x^2} \right\} \sqrt{1-x^2} -$$
$$\qquad \frac{1\cdot 3\cdot 5\cdots\cdots(m-2)}{2\cdot 4\cdot 6\cdots\cdots(m-1)}\log. \frac{1+\sqrt{1-x^2}}{x} + C$$

If we put m successively equal to 0, 2, 4, &c. we have

$$\int \frac{dx}{\sqrt{1-x^2}} = \sin.^{-1} x + C$$

$$\int \frac{dx}{x^2\sqrt{1-x^2}} = - \frac{\sqrt{1-x^2}}{x} + C$$

$$\int \frac{dx}{x^4 \sqrt{1-x^2}} = -\frac{\sqrt{1-x^2}}{3x^3} + \frac{2}{3}\int \frac{dx}{x^2 \sqrt{1-x^2}}$$

$$\int \frac{dx}{x^6 \sqrt{1-x^2}} = -\frac{\sqrt{1-x^2}}{5x^5} + \frac{4}{5}\int \frac{dx}{x^4 \sqrt{1-x^2}}$$

$$\vdots$$

$$\int \frac{dx}{x^m \sqrt{1-x^2}} = -\frac{\sqrt{1-x^2}}{(m-1)x^{m-1}} + \frac{m-2}{m-1}\int \frac{dx}{x^{m-2}\sqrt{1-x^2}}$$

or, by substituting,

$$\int \frac{dx}{\sqrt{1-x^2}} = \sin.^{-1} x + C$$

$$\int \frac{dx}{x^2 \sqrt{1-x^2}} = -\frac{\sqrt{1-x^2}}{x} + C$$

$$\int \frac{dx}{x^4 \sqrt{1-x^2}} = -\left(\frac{1}{3x^3} + \frac{2}{1\cdot 3x}\right)\sqrt{1-x^2} + C$$

$$\int \frac{dx}{x^6 \sqrt{1-x^2}} = -\left(\frac{1}{5x^5} + \frac{4}{3\cdot 5x^3} + \frac{2\cdot 4}{1\cdot 3\cdot 5\, x}\right)\sqrt{1-x^2} + C$$

$$\vdots$$

$$\int \frac{dx}{x^m\sqrt{1-x^2}} = -\left\{\frac{1}{(m-1)x^{m-1}} + \frac{m-2}{(m-3)(m-1)x^{m-3}} + \right.$$

$$\frac{(m-4)(m-2)}{(m-5)(m-3)(m-1)x^{m-5}}$$

$$\left.\cdots\cdots\cdots \frac{1\cdot 2\cdot 4\cdots(m-2)}{1\cdot 3\cdot 5\cdots(m-1)\,x}\right\}\sqrt{1-x^2} + C.$$

5. To integrate

$$\frac{x^m dx}{\sqrt{2ax-x^2}}.$$

This expression is the same as

$$\frac{x^{m-\frac{1}{2}}dx}{\sqrt{2a-x}}.$$

and, comparing this with the formula V. we have, for m, a, b, n, and p, $m-\frac{1}{2}$, $2a$, -1, 1 and $-\frac{1}{2}$, therefore

$$\int \frac{x^m dx}{\sqrt{2ax - x^2}} = - \frac{x^{m-1}\sqrt{2ax - x^2}}{m} + \frac{a\,(2m - 1)}{m} \int \frac{x^{m-1}\,dx}{\sqrt{2ax - x^2}},$$

so that, by continuing thus to diminish the exponent m, the integral of the proposed differential will finally depend upon

$$\int \frac{dx}{\sqrt{2ax - x^2}} = \text{versin.}^{-1} \frac{1}{a}\, x + \text{C.}$$

6. To integrate

$$\frac{x^m\, dx}{(a^2 + x^2)^p}.$$

If m is greater than, or equal to, 2, this differential may be reduced by formula IV. or V. or by the application of both to the forms

$$\frac{dx}{(a^2 + x^2)^{p'}} \text{ or } \frac{xdx}{(a^2 + x^2)^{p'}},$$

according as m is even or odd. The integral of the second form is

$$\frac{1}{2\,(1 - p)\,(a^2 + x^2)^{p'-1}} + \text{C};$$

but the first form is not generally integrable, unless p' is $\frac{1}{2}$, or some multiple of it, in which case it may be further reduced by formula III. and will finally depend upon

$$\int \frac{dx}{\sqrt{a^2 - x^2}} = \log. \text{ C } (x + \sqrt{a^2 + x^2}).$$

If $m = 0$ and p be a whole number, formula III. gives

$$\int \frac{dx}{(a^2 + x^2)^p} = \frac{1}{2\,(p - 1)\,a^2} \cdot \frac{x}{(a^2 + x^2)^{p-1}}$$
$$+ \frac{2p - 3}{2\,(p - 1)\,a^2} \int \frac{dx}{(a^2 + x^2)^{p-1}},$$

which is the equation otherwise deduced in art. (13),

7. To integrate

$$\frac{dx}{x^4 \sqrt{a + bx^2}}.$$

$$\int \frac{dx}{x^4 \sqrt{a + bx^2}} = \left(- \frac{1}{3ax^3} + \frac{2b}{3a^2x} \right) \sqrt{a + bx^2} + \text{C.}$$

8. To integrate

$$\frac{dx}{(a + bx^2)^{\frac{p}{2}}}$$

for the odd values of p.

$$\int \frac{dx}{(a + bx^2)^{\frac{1}{2}}} = \frac{1}{\sqrt{b}} \log. \{x \sqrt{b} + \sqrt{a + bx^2}\} + C. \quad \text{(See ex. 7, p.34)}$$

$$\int \frac{dx}{(a + bx^2)^{\frac{3}{2}}} = \frac{x}{a \sqrt{a + bx^2}} + C$$

$$\int \frac{dx}{(a + bx^2)^{\frac{5}{2}}} = \left\{ \frac{1}{3a (a + bx^2)} + \frac{2}{3a^2} \right\} \frac{x}{\sqrt{a + bx^2}} + C$$

$$\int \frac{dx}{(a + bx^2)^{\frac{7}{2}}} = \left\{ \frac{1}{5a (a + bx^2)^2} + \frac{4}{15 a^2 (a + bx^2)} + \frac{1}{15 a^3} \right\} \frac{x}{\sqrt{a + bx^2}} + C$$

&c.   &c.

9. To integrate

$$\frac{x^m dx}{\sqrt{a + bx^2}}$$

for the odd values of m.

$$\int \frac{x dx}{\sqrt{a + bx^2}} = \frac{\sqrt{a + bx^2}}{b} + C$$

$$\int \frac{x^3 dx}{\sqrt{a + bx^2}} = \left\{ \frac{x^2}{3b} - \frac{2a}{3b^2} \right\} \sqrt{a + bx^2} + C$$

$$\int \frac{x^5 dx}{\sqrt{a + bx^2}} = \left\{ \frac{x^4}{5b} - \frac{4ax^2}{15b^2} + \frac{8a^2}{15b^3} \right\} \sqrt{a + bx^2} + C.$$

&c.   &c.

10. To integrate

$$\frac{x^m dx}{\sqrt{a + bx^2}}$$

for the even values of m.

$$\int \frac{dx}{\sqrt{a + bx^2}} = \frac{1}{\sqrt{b}} \log. \{x \sqrt{b} + \sqrt{a + bx^2}\} = L$$

$$\int \frac{x^2 dx}{\sqrt{a + bx^2}} = \frac{x \sqrt{a + bx^2}}{2b} - \frac{a}{2b} L + C$$

$$\int \frac{x^4 dx}{\sqrt{a + bx^2}} = \{\frac{x^3}{4b} - \frac{3ax}{8b^2}\} \sqrt{a + bx^2} + \frac{3a^2}{8b^2} L + C$$

$$\&c. \qquad\qquad \&c.$$

11. To integrate

$$\frac{\sqrt{a + bx^2}}{x^m} dx$$

for odd values of m.

$$\int \frac{\sqrt{a + bx^2}}{x} dx = \sqrt{a + bx^2} + \sqrt{a} \log. \frac{\sqrt{a + bx^2} - \sqrt{a}}{x} + C$$

$$\int \frac{\sqrt{a + bx^2}}{x^3} dx = - \frac{\sqrt{a + bx^2}}{2x^2} + \frac{b}{2\sqrt{a}} \log. \frac{\sqrt{a + bx^2} - \sqrt{a}}{x} + C$$

$$\int \frac{\sqrt{a + bx^2}}{x^3} dx = - \{\frac{a + bx^2}{4ax^4} - \frac{b}{8ax^2}\} \sqrt{a + bx^2}$$

$$- \frac{b^2}{8a^{\frac{3}{2}}} \log. \frac{\sqrt{a + bx^2} - \sqrt{a}}{x} + C.$$

$$\&c. \qquad\qquad \&c.$$

When the proposed binomial cannot be reduced to a form integrable by the preceding methods, then the only general mode of procedure is to develope the binomial in a series, and to integrate each term separately. The method of integration by series will be treated of in a future chapter.

CHAPTER IV.

ON THE INTEGRATION OF LOGARITHMIC AND EXPONENTIAL FORMS.

Logarithmic Forms.

(26.) But few of these forms are capable of integration by any general process at present known, except, indeed, by the method of series, which furnishes, however, but an approximation, and should therefore be resorted to only when exact methods fail.

To integrate the form

$$X \log.^n x dx,$$

in which X is a function of x.

If, in the formula for integration by parts, viz.

$$\int u dv = uv - \int v du,$$

we suppose

$$dv = X dx, \ u = \log.^n x,$$

we have

$$\int X dx \log.^n x = \log.^n x \int X dx - \int \left(\int X dx \cdot n \log.^{n-1} x \frac{dx}{x} \right),$$

or putting, for brevity,

$$\int X dx = X_1$$

$$\int X dx \log.^n x = \log.^n x \cdot X_1 - n \int \frac{X_1}{x} \log.^{n-1} x dx \ \cdot \cdot \cdot \cdot (1).$$

If n is a positive whole number, the successive application of this formula will finally reduce the integration of the proposed form to that of an algebraic function, so that the proposed will be integrable, provided we can integrate, in succession, the algebraic functions

$$X dx = dX_1, \ \frac{X_1}{x} dx = dX_2, \ \frac{X_2}{x} dx = dX_3, \ \&c.$$

To give an example of the application of this formula, suppose we had to integrate

$$\frac{x \log.x\, dx}{\sqrt{a^2 + x^2}}.$$

Here

$$n = 1, \ \mathrm{X} = \frac{x}{\sqrt{a^2 + x^2}} \ \therefore \ \mathrm{X} dx = d\mathrm{X}_1 = \frac{x\, dx}{\sqrt{a^2 + x^2}} \ \therefore \ \mathrm{X}_1 = \sqrt{a^2 + x^2},$$

consequently the final integral will be

$$\mathrm{X}_2 = \int \frac{\sqrt{a^2 + x^2}}{x}\, dx,$$

which we may at once reduce by the formula I. last chapter; or, if we multiply numerator and denominator of this by the numerator, we have

$$\int \frac{a^2 dx}{x \sqrt{a^2 + x^2}} + \int \frac{x\, dx}{\sqrt{a^2 + x^2}} = \int \frac{a^2 dx}{x \sqrt{a^2 + x^2}} + \sqrt{a^2 + x^2},$$

and, (ex. 8, p. 34,)

$$a^2 \int \frac{dx}{x \sqrt{a^2 + x^2}} = - a \ \log. \frac{\sqrt{a^2 + x^2} + a}{x} + \mathrm{C}.$$

consequently, by the formula (1),

$$\int \frac{x \log. x\, dx}{\sqrt{a^2 + x^2}} = \log. x \ . \ \sqrt{a^2 + x^2} - \sqrt{a^2 + x^2} +$$

$$a \ \log. \frac{\sqrt{a^2 + x^2} + a}{x} + \mathrm{C}.$$

(27.) One of the most useful cases of the above general form is that in which $\mathrm{X} = x^m$, the form then being

$$x^m \log.^n x\, dx,$$

and for which the formula of reduction (1) is

$$\int x^m\, dx \log.^n x = \frac{x^{m+1}}{m + 1} \log.^n x - \frac{n}{m + 1} \int x^m \log.^{n-1} x\, dx \ \ldots \ (2),$$

$$\therefore \int x^m\, dx \log.^{n-1} x = \frac{x^{m+1}}{m + 1} \log.^{n-1} x - \frac{n - 1}{m + 1} \int x^m \log.^{n-2} x\, dx$$

$$\int x^m\, dx \log.^{n-2} x = \frac{x^{m+1}}{m + 1} \log.^{n-2} x - \frac{n - 2}{m + 1} \int x^m \log.^{n-3} dx$$

$$\int x^m \, dx \, \log.^{n-3} x = \frac{x^{m+1}}{m+1}\log.^{n-3}x - \frac{n-3}{m+1}\int x^m \log.^{n-4} x \, dx$$

&c.   &c.

Hence, substituting for the integrals, on the right, their values as given by the succeeding equations, we have generally

$$\int x^m \, dx \log.^n x = \frac{x^{m+1}}{m+1}\left\{\log.^n x - \frac{n}{m+1}\log.^{n-1}x + \frac{n(n-1)}{(m+1)^2}\log.^{n-2}x\right.$$
$$\left. - \frac{n(n-1)(n-2)}{(m+1)^2}\log.^{n-3}x + \&c.\right\} + C \cdot \cdot \cdot \cdot \cdot (3).$$

This series terminates whenever n is a positive integer. It fails to be applicable, however, if $m = -1$, in which case the differential is

$$\log.^n x \cdot \frac{dx}{x} = \log.^n x \cdot d \log. x$$

$$\therefore \int \log.^n x \cdot \frac{dx}{x} = \frac{\log.^{n+1} x}{n+1} + C,$$

so that, in this case, the formula is not required.

This last expression, if n is negative, becomes

$$-\int \frac{dx}{x \log.^n x} = \frac{\log.^{-n+1} x}{-n+1} \, ;$$

so that, calling this v and x^{m+1}, u, the formula for the integration by parts gives

$$\int \frac{x^m \, dx}{\log.^n x} = -\frac{x^{m+1}}{(n-1)\log.^{n-1} x} + \frac{m+1}{n-1}\int \frac{x^m \, dx}{\log.^{n-1} x} \cdot \cdot \cdot \cdot (4),$$

or, proceeding as in the former case,

$$\int \frac{x^m \, dx}{\log^n x} = -\frac{x^{m+1}}{n-1}\left\{\frac{1}{\log.^{n-1} x} + \frac{m+1}{n-2}\cdot\frac{1}{\log.^{n-2} x} + \right.$$
$$\left. \frac{(m+1)^2}{(n-2)(n-3)}\cdot\frac{1}{\log.^{n-3} x} + \cdot\cdot\cdot\cdot\right\} +$$
$$\frac{(m+1)^{n-1}}{1\cdot 2\cdot 3\cdot\cdot\cdot\cdot(n-1)}\int \frac{x^m \, dx}{\log. x} \cdot \cdot \cdot \cdot \cdot (5);$$

beyond the integral $\int \dfrac{x^m dx}{\log. x}$ the reduction cannot be carried, for the formula ceases to be applicable when n becomes $=1$. This final integral may be put in a somewhat simpler form by substituting z for x^{m+1}, for then $x^m\, dx = \dfrac{dx}{m+1}$, and log. $x = \dfrac{\log. z}{m+1}$, consequently

$$\int \frac{x^m\, dx}{\log. x} = \int \frac{dz}{\log. z},$$

which expression, simple as it is in appearance, has never yet been integrated except by series.

When n is a fraction either positive or negative, we may, by means of one or other of these formulas, reduce the integration to that of another expression of the same form, in which n will be comprised between 1 and -1, which final expression must then be integrated by series.

EXAMPLES.

(28.) 1. Required the integral of $x^3\, dx \log.^2 x$.

Since here $m = 3$ and $n = 2$, the formula (3) becomes

$$\int x^3\, dx \log.^2 x = \frac{x^4}{4} \left\{ \log.^2 x - \frac{1}{2} \log. x + \frac{1}{4} \right\} + C.$$

2. Required the integral of

$$\frac{x^4\, dx}{\log.^2 x}.$$

The formula (5) gives

$$\int \frac{x^4\, dx}{\log.^2 x} = - \frac{x^5}{\log. x} + 5 \int \frac{x^4\, dx}{\log. x}.$$

This last integral, as before observed, cannot be obtained in finite terms; but, if we put z for x^{m+1}, the form, as before shewn, becomes $\int \dfrac{dz}{\log. z}$. Now, if log. z be u, then (*Diff. Calc.* p. 29),

$$z = e^u = 1 + u + \frac{u^2}{2} + \frac{u^3}{2 \cdot 3} + \&c.$$

consequently

$$\int \frac{dz}{\log. z} = \int \frac{e^u\, du}{u} = \int \frac{du}{u} + \int du + \frac{1}{2} \int u\, du +$$

$$+ \frac{1}{2 \cdot 3} \int u^2 \, du + \&c.$$

$$= C + \log. u + u + \frac{u^2}{2^2} + \frac{u^3}{2 \cdot 3^2} + \&c.$$

$$= C + (\log.)^2 z + \log. z + \frac{\log.^2 z}{2^2} + \frac{\log.^3 z}{2 \cdot 3^2} + \&c.$$

$$\therefore \int \frac{x^4 \, dx}{\log. x} = C + (\log.)^2 x^5 + \log. x^5 + \frac{\log.^2 x^5}{2^2} + \frac{\log.^3 x^5}{2 \cdot 3^2} + \&c.$$

3. To integrate

$$\frac{dx}{x \log.^n x}.$$

$$\int \frac{dx}{x \log.^n x} = C - \frac{1}{(n-1) \log.^{n-1} x}.$$

4. To integrate

$$\frac{x^4 \, dx}{\log.^3 x}.$$

$$\int \frac{x^4 \, dx}{\log.^3 x} = - \frac{x^5}{2 \log.^2 x} - \frac{5 x^5}{2 \log. x} +$$

$$\frac{25}{2} \left\{ (\log.)^2 x^5 + \log. x^5 + \frac{\log.^2 x^5}{2^2} + \frac{\log.^3 x^5}{2 \cdot 3^2} + \&c. \right\} + C.$$

(29.) It may be here remarked that the formula (2) is rather more comprehensive than it appears to be, for, by attending to the manner in which it has been deduced, we readily perceive that it equally holds, when instead of x^m, we substitute $(x \pm a)^m$, so that the integral of

$$(x \pm a)^m \, dx \, \log.^n x$$

will be given by the right hand member of (3), provided that in the factor without the brackets we change x into $x \pm a$.

The same is true of the expression (5), although we cannot legitimately infer this from the manner in which we have deduced the formula (4). If, however, as in the first case, we commence with the more general form $\dfrac{X \, dx}{\log.^n x}$, which may be written $X x \cdot \dfrac{dx}{x \log.^n x}$, then, since

$$\int \frac{dx}{x \log.^n x} = \int \frac{dx \log.^{-n} x}{x} = \frac{\log.^{-n+1} x}{-n+1} = -\frac{1}{(n-1) \log.^{n-1} x},$$

we shall have, by integrating by parts,

$$\int \frac{X dx}{\log.^n x} = -\frac{X x}{(n-1) \log.^{n-1} x} + \frac{1}{n-1} \int \frac{d(Xx)}{\log.^{n-1} x},$$

or, if $X = x^m$,

$$\int \frac{x^m dx}{\log.^n x} = -\frac{x^{m+1}}{(n-1) \log.^{n-1} x} + \frac{m+1}{n-1} \int \frac{x^m dx}{\log.^{n-1} x},$$

in which obviously $(x \pm a)^m$ may be put for x^m.

Exponential forms.

(30.) Let us now consider exponential forms; these, like logarithmic, are for the most part unintegrable exactly.

To integrate the form.

$$a^x \cdot x^m \, dx.$$

Putting in the formula for integration by parts

$$u = x^m, \; dv = a^x dx \therefore v = \frac{a^x}{\log.a},$$

it becomes

$$\int x^m a^x dx = \frac{x^m a^x}{\log. a} - \frac{m}{\log. a} \int x^{m-1} a^x dx$$

$$\therefore \int x^{m-1} a^x dx = \frac{x^{m-1} a^x}{\log. a} - \frac{m-1}{\log. a} \int x^{m-2} a^x dx$$

$$\int x^{m-2} a^x dx = \frac{x^{m-2} a^x}{\log. a} - \frac{m-2}{\log. a} \int x^{m-3} a^x dx,$$

    &c.   &c.

Consequently by substitution,

$$\int x^m a^x dx = \frac{a^x}{\log. a} \left\{ x^m - \frac{m x^{m-1}}{\log. a} + \frac{m(m-1) x^{m-2}}{\log.^2 a} - \dots \right.$$

$$\left. \pm \frac{1 \cdot 2 \cdot 3 \dots m}{\log.^m a} \right\} + C,$$

the upper sign of the last term having place when m is even, and the lower when m is odd.

When m is negative, the series within the brackets does not terminate, and is therefore inapplicable; but if in this case we put

$$u = a^x, \ dv = x^{-m} \, dx \ \therefore \ v = \frac{-1}{(m-1)\,x^{m-1}},$$

the formula for integration by parts will give

$$\int \frac{a^x \, dx}{x^m} = -\frac{a^x}{(m-1)\,x^{m-1}} + \frac{\log. a}{m-1} \int \frac{a^x \, dx}{x^{m-1}}$$

$$\therefore \int \frac{a^x \, dx}{x^{m-1}} = -\frac{a^x}{(m-2)\,x^{m-2}} + \frac{\log. a}{m-2} \int \frac{a^x \, dx}{x^{m-2}}$$

$$\int \frac{a^x \, dx}{x^{m-2}} = -\frac{a^x}{(m-3)\,x^{m-2}} + \frac{\log. a}{m-2} \int \frac{a^x \, dx}{x^{m-3}}$$

therefore, by substitution,

$$\int \frac{a^x \, dx}{x^m} = -\frac{a^x}{(m-1)\,x^{m-1}} \left\{ 1 + \frac{\log. a}{m-2} x + \frac{\log.^2 a}{(m-2)(m-3)} x^2 + \right.$$

$$\left. \dots + \frac{\log.^{m-2} a}{(m-2)(m-3)\dots 1} x^{m-2} \right\} + \frac{\log.^{m-1} a}{1 \cdot 2 \cdot 3 \dots (m-1)} \int \frac{a^x \, dx}{x}.$$

The integral $\displaystyle\int \frac{a^x \, dx}{x}$ is not rigorously determinable, but it may be approximated to by series.

EXAMPLES.

(31.)   1. To integrate

$$\frac{a^x \, dx}{x^3}.$$

The formula just deduced gives

$$\int \frac{a^x \, dx}{x^3} = -\frac{a^x}{2x^2} \cdot \{1 + \log. a \cdot x\} + \frac{\log.^2 a}{2} \int \frac{a^x \, dx}{x},$$

but if we substitute for a^x its development (*Diff. Calc.* p. 29), we have

$$\frac{a^x \, dx}{x} = \frac{dx}{x} + \log. a \, dx + \left(\frac{1}{2} \log.^2 a \cdot x + \frac{1}{2 \cdot 3} \log.^3 a \cdot x^2 + \&c. \right) dx$$

$$\therefore \int \frac{a^x\, dx}{x} = \log. x + \log. a . x + \frac{1}{2} \log.^2 a \cdot \frac{x^2}{1\cdot 2} +$$

$$\frac{1}{3} \log.^3 a \cdot \frac{x^3}{1\cdot 2\cdot 3} + \&c.$$

Hence

$$\int \frac{a^x\, dx}{x^3} = -\frac{a^x}{2x^2}\{1 + \log. a . x\} +$$

$$\frac{\log.^2 a}{2}\{\log. x + \log. a\cdot x + \frac{1}{2}\log.^2 a \cdot \frac{x^2}{1\cdot 2} + \&c.\} + C.$$

2. To integrate

$$x^3 a^x\, dx.$$

$$\int x^3 a^x\, dx = \frac{a^x}{\log. a}\{x^3 - \frac{3x^2}{\log. a} + \frac{6x}{\log.^2 a} - \frac{6}{\log.^3 a}\} + C.$$

3. To integrate

$$\frac{e^x \cdot x\, dx}{(1+x)^2}.*$$

$$\int \frac{e^x\, x\, dx}{(1+x)^2} = \frac{e^x}{1+x} + C.$$

4. To integrate

$$\frac{a^x \cdot x\, dx}{(b+x)^2}.$$

$$\int \frac{a^x \cdot x\, dx}{(b+x)^2} = \frac{1}{a^b}\{\frac{b a^y}{y} + (1 - b \log. a)\int \frac{a^y\, dy}{y}\},$$

where $y = b + a$.

5. To determine a general formula for the integration of

$$a^{-x} x^m\, dx$$

$$\int a^{-x} x^m dx = -\frac{a^{-x}}{\log. a}\{x^m + \frac{m x^{m-1}}{\log. a} + \frac{m\,(m-1)\,x^{m-2}}{\log.^2 a} +$$

$$\ldots + \frac{1\cdot 2\cdot 3 \ldots m}{\log.^m a}\} + C.$$

---

* By putting $1 + x = Y$, this will be transformed into

$$\frac{1}{e}\{\frac{e^y\, dy}{y} - \frac{e^y\, dy}{y^3}\}.$$

CHAPTER V.

ON THE INTEGRATION OF TRIGONOMETRICAL AND CIRCULAR FUNCTIONS.

(32.) In considering the differential expressions whose coefficients are functions of trigonometrical lines, it is obvious that we may confine our attention to those only which contain sines and cosines, since all the other lines may be converted into functions of these. As in the former chapter, so here, we shall treat of those forms only to which general processes apply, omitting all notice of the almost infinite variety of combinations which might be devised, and for which the calculus in its present state supplies us with no rule of integration.

(33.) *To integrate the form*

$$\sin.^{m} x \cos.^{n} x \, dx.$$

To this general expression we may apply the method of integration by parts, first putting it under the more convenient form

$$\sin.^{m-1} x \cos.^{n} x \sin. x \, dx,$$

for, comparing this with the formula

$$\int u dv = uv - \int v du,$$

by assuming

$$\sin.^{m-1} x = u, \quad \cos.^{n} x \sin. x \, dx = -\cos.^{n} x \, d \cos. x = dv,$$

and therefore

$$(m-1) \sin.^{m-2} x \cos. x \, dx = du, \quad -\frac{\cos.^{n+1} x}{n+1} = v,$$

it becomes

$$\int \sin.^m x \cos.^n x \, dx = -\frac{\sin.^{m-1} x \cos.^{n+1} x}{\cdot n + 1}$$

$$+ \frac{m-1}{n+1} \int \sin.^{m-2} x \cos.^{n+2} x \, dx.$$

This form may be somewhat simplified, for, by substituting in the integral on the right

$$\cos.^n x \, (1 - \sin.^2 x) \text{ for } \cos.^{n+2} x$$

it becomes divisible into the two

$$\int \sin.^{m-2} x \cos.^n x \, dx - \int \sin.^m x \cos.^n x \, dx;$$

making, therefore, this substitution, we obtain from the result

$$\int \sin.^m x \cos.^n x \, dx = -\frac{\sin.^{m-1} x \cos.^{n+1} x}{m + n}$$

$$+ \frac{m-1}{m+n} \int \sin.^{m-2} x \cos.^n x \, dx$$

$$\therefore \int \sin.^{m-2} x \cos.^n x \, dx = -\frac{\sin.^{m-3} x \cos.^{n+1} x}{m - 2 + n}$$

$$+ \frac{m-3}{m-2+n} \int \sin.^{m-4} x \cos.^n x \, dx$$

$$\int \sin.^{m-4} x \cos.^n x \, dx = -\frac{\sin.^{m-5} x \cos.^{n+1} x}{m - 4 + n}$$

$$+ \frac{m-5}{m-4+n} \int \sin.^{m-6} x \cos.^n x \, dx,$$

&c.   &c.

Hence since the exponent m is thus diminished by 2 at each successive application of this formula, while the exponent n remains the same, it follows, that if m is a positive odd number, m and n being also both integers, the integration will be finally reduced to

$$\int \sin. x \cos.^n x \, dx = -\int \cos.^n x \, d \sin. x = -\frac{\cos.^{n+1} x}{n+1} + C,$$

so that, in this case, the proposed may be completely integrated by the application of this formula of reduction.

(34.) If we substitute for the integrals in the right hand members of the above equations their values as given by the succeeding equations, we shall have the following

General Formulas for the Reduction of the Integral

$$\int \sin.^m x \cos.^n x \, dx.$$

I.

$$\int \sin.^m x \cos.^n x \, dx =$$

$$-\frac{\cos.^{n+1} x}{m+n}\left\{ \sin.^{m-1}x + \frac{m-1}{m-2+n}\sin.^{m-3}x + \frac{(m-1)(m-3)}{(m-2+n)(m-4+n)}\sin.^{m-5}x + \ldots \right\} + \frac{(m-1)(m-3)(m-5)\ldots}{(m+n)(m-2+n)(m-4+n)\ldots} \left\{ \begin{array}{c} \int \sin.x\cos.^n x\,dx \\ \text{or} \\ \int \cos.^n x \, dx \end{array} \right.$$

the upper being the final integral, if, m is dd, and the l wr, if m is een.

In the application of this formula to particular cases it must be remembered that the terms within the brackets are to be cntinued as far as the exponents tne greater than 0, but no farther; the 1 nont at wih we stop will be the final factor of the rtor of the term whut the brackets.

If for a we substitute its complement $\dfrac{\pi}{2} - x$, and interchange m and n, the formula becomes

II.

$$\int \sin.^m x \cos.^n x \, dx =$$

$$\frac{\sin.^{m+1} x}{n+m}\left\{ \cos.^{n-1} x + \frac{n-1}{n-2+m}\cos.^{n-3}x + \frac{(n-1)(n-3)}{(n-2+m)(n-4+m)}\cos.^{n-5}x + \ldots \right\} + \frac{(n-1)(n-3)(n-5)\ldots}{(n+m)(n-2+m)(n-4+m)\ldots} \left\{ \begin{array}{c} \int \cos.x\sin.^m x\,dx \\ \text{or} \\ \int \sin.^m x \, dx \end{array} \right.$$

the upper being the final integral if n is odd, and the lower, if n is even.

The terms within the brackets are to be continued as far as the positive exponents extend, as in formula I. and the exponents at which we stop will be the last factor in the numerator of the term without the brackets.

Let now $n = 0$, in formula I., then

III.

$$\int \sin.^m x\, dx = -\frac{\cos. x}{m} \left\{ \sin.^{m-1} x + \frac{m-1}{m-2} \sin.^{m-3} x + \frac{(m-1)(m-3)}{(m-2)(m-4)} \sin.^{m-5} x + \ldots \right\} + \frac{(m-1)(m-3)(m-5)\ldots}{m\,(m-2)\,(m-4)\ldots} \left\{ \begin{array}{c} \int \sin. x\, dx \\ \text{or} \\ \int dx \end{array} \right.$$

the upper being the final integral if m is odd, and the lower if m is even.

In like manner, putting $m = 0$, in formula II., we have

IV.

$$\int \cos.^n x\, dx = \frac{\sin. x}{n} \left\{ \cos.^{n-1} x + \frac{n-1}{n-2} \cos.^{n-3} x + \frac{(n-1)(n-3)}{(n-2)(n-4)} \cos.^{n-5} x + \ldots \right\} + \frac{(n-1)(n-3)(n-5)\ldots}{n\,(n-2)\,(n-4)\ldots} \left\{ \begin{array}{c} \int \cos. x\, dx \\ \text{or} \\ \int dx \end{array} \right.$$

where, as before, the upper integral is to be used for n odd, and the lower for n even. By the joint application of these formulas the proposed form, when m and n are positive integers, is therefore reducible to one or other of the forms

$$\int \sin. x \cos.^n x\, dx = -\frac{\cos.^{n+1} x}{n+1} + C, \quad \int \cos. x \sin.^m x\, dx = \frac{\sin.^{m+1} x}{m+1} + C, \quad \int \sin. x\, dx = C - \cos. x, \quad \int \cos. x\, dx = \sin. x + C,$$

$$\int dx = x + C.$$

Before considering the case in which m, n, are one or both negative, we shall give an example or two of the application of the preceding formulas.

EXAMPLES.

(35.) 1. To integrate

$$\sin.^4 x \cos.^3 x \, dx.$$

As, in this expression, n is odd, we shall employ the formula II., which gives

$$\sin.^4 x \cos.^3 x \, dx = \frac{\sin.^5 x}{7} \cos.^2 x + \frac{2}{7} \cdot \frac{\sin.^5 x}{5} + C$$

$$= \frac{\sin.^5 x}{7} (1 - \sin.^2 x) + \frac{2 \sin.^5 x}{7 \cdot 5} + C$$

$$= \frac{1}{5} \sin.^5 x - \frac{1}{7} \sin.^7 x + C.$$

2. To integrate

$$\sin.^5 x \cos. x^4 \, dx.$$

As m is odd it will be best to employ formula I. which gives

$$\int \sin.^5 x \cos. x^4 \, dx = - \frac{\cos.^5 x}{9} \left\{ \sin.^4 x + \frac{4}{7} \sin.^2 x \right\} +$$

$$\frac{4 \cdot 2}{9 \cdot 7} \cdot \frac{- \cos.^5 x}{5} + C$$

$$= - \frac{\cos.^5 x}{9} \left\{ \sin.^4 x + \frac{4}{7} \sin.^2 x + \frac{4 \cdot 2}{7 \cdot 5} \right\} + C.$$

3. To integrate

$$\sin.^3 x \, dx.$$

By formula III.

$$\int \sin.^3 dx = - \frac{\cos. x}{3} \sin.^2 x \qquad \frac{2}{3} \cos. x + C$$

$$= - \frac{1}{3} \cos. x (\sin.^2 x + 2) + C.$$

4. To integrate

$$\sin.^3 x \cos.^2 x \, dx.$$

$$\int \sin.^3 x \cos.^2 x \, dx = -\frac{\cos.^3 x}{5}\left(\sin.^2 x + \frac{2}{3}\right) + C.$$

5. To integrate

$$\sin.^6 x \, dx.$$

$$\int \sin.^6 x \, dx = -\frac{\cos. x}{6}.\left\{\sin.^5 x + \frac{5}{4}\sin.^3 x + \frac{5 \cdot 3}{4 \cdot 2}\sin. x\right\} +$$

$$\frac{5 \cdot 3 \cdot 1}{6 \cdot 4 \cdot 2} x + C.$$

6. To integrate

$$\sin.^8 x \cos.^3 x \, dx$$

$$\int \sin.^8 x \cos.^3 x \, dx = \frac{\sin.^9 x}{11}\left(\cos.^2 x + \frac{2}{9}\right) + C.$$

It is worth while to observe here, that when either of the exponents m, n, is 3, the formula for the integral is so remarkably simple that in every such case the integral may be instantly written down without any reference to the table at page 62. For by formula I.

$$\int \sin.^3 x \cos.^n x \, dx = -\frac{\cos.^{n+1} x}{n+3}\left\{\sin.^2 x + \frac{2}{n+1}\right\} + C,$$

and by formula II.

$$\int \sin.^m x \cos.^3 x \, dx = \frac{\sin.^{m+1} x}{m+3}\left\{\cos.^2 x + \frac{2}{m+1}\right\} + C,$$

which two forms may be remembered and applied without any trouble.

Let us now suppose that one of the exponents m, n, is negative, we shall then have

(36). *To integrate the forms*

$$\frac{\sin.^m x}{\cos.^n x} dx, \quad \frac{\cos.^n x}{\sin.^m x} dx.$$

The formulas hitherto given will not suffice for this purpose; they might, indeed, by means of the formulas I. and II., be reduced to the forms

$$\frac{\sin. x}{\cos.^n x}\, dx \text{ or } \frac{xdx}{\cos.^n x}, \text{ and } \frac{\cos. x}{\sin.^m x}\, dx \text{ or } \frac{xdx}{\sin.^m x},$$

of which the two

$$\frac{\sin. x}{\cos.^n x}\, dx \text{ and } \frac{\cos. x}{\sin.^m x}dx$$

are immediately integrable, so that, when m is odd, in the first of the above forms, and when n is odd, in the second, the formulas I. and II. respectively apply. When, however, this is not the case, we are led to the forms

$$\frac{xdx}{\sin.^m x},\ \frac{xdx}{\cos.^n x},$$

which have not as yet been integrated, for m, n in the formulas III, IV. are essentially positive; the question is therefore reduced to the integration of these two forms. Taking the first we have

$$\int \frac{xdx}{\sin.^m x} = -\int \sin.^{-m-1} x\, d\cos.x \ \ldots \ldots \ (1),$$

and, putting

$$\sin.^{-m-1} x = u,\ d\cos. x = dv,$$

the formula

$$-\int udv = -uv + \int vdu$$

becomes

$$-\int \sin.^{-m-1} x\, d\cos. x = -\sin.^{-m-1} x\, d\cos. x -$$

$$(n+1)\int \sin.^{-m-2} x\, \cos.^2 dx$$

$$= -\sin.^{-m-1} x\, \cos. x - (m+1)\left\{\int \sin.^{-m-2} x - \int \sin.^{-m} x\right\} dx,$$

or, dividing by $m+1$, and transposing, we have, in virtue of (1),

$$\int \frac{dx}{\sin^{m+2} x} = -\frac{\cos. x}{(m+1)\sin.^{m+1}x} + \frac{m}{m+1}\int \frac{dx}{\sin.^m x},$$

or, putting m for $m+2$,

$$\int \frac{dx}{\sin.^m x} = -\frac{\cos. x}{(m-1)\sin.^{m-1} x} + \frac{m-2}{m-1} \int \frac{dx}{\sin.^{m-2} x} \quad ....(2)$$

$$\therefore \int \frac{dx}{\sin.^{m-2} x} = -\frac{\cos. x}{(m-3)\sin.^{m-3} x} + \frac{m-4}{m-3} \int \frac{dx}{\sin.^{m-4} x}$$

$$\int \frac{dx}{\sin.^{m-4} x} = -\frac{\cos. x}{(m-5)\sin.^{m-5} x} + \frac{m-6}{m-5} \int \frac{dx}{\sin.^{m-6} x}.$$

&c. &c.

By the application of this process, we see that when m is odd, the integration will be finally reduced to that of $\int \dfrac{dx}{\sin. x}$ which, by multiplying numerator and denominator by sin. x, becomes

$$\int \frac{dx}{\sin. x} = \frac{\sin. x \, dx}{1 - \cos.^2 x} = -\frac{d\cos. x}{1 - \cos.^2 x} =$$

$$\text{(see ex. 1, p. 14), log. } \left(\frac{1 - \cos. x}{1 - \cos. x}\right)^{\frac{1}{2}} + C,$$

that is, (*Dr. Gregory's Trigonometry, page 47,*)

$$\int \frac{dx}{\sin. x} = \log. \tan. \frac{1}{2} x + C,$$

when m is even, the final integral is simply $\int dx = x + C$, or we may, in this case, stop at the preceding integral, which will be

$$\int \frac{dx}{\sin.^2 a} = \cot. x + C.$$

(37.) By substituting for the several integrals on the right their values as given by the succeeding equations, we shall have the following continuation of the table of formulas given at page 62:

General Formulas for the Determination of the Integrals

$$\int \frac{dx}{\sin.^m x} \quad \text{and} \quad \int \frac{dx}{\cos.^n x}.$$

V.

$$\int \frac{dx}{\sin.^m x} =$$

$$-\frac{\cos. x}{m-1}\left\{\frac{1}{\sin.^{m-1} x}+\frac{m-2}{(m-3)\sin.^{m-3} x}+\frac{(m-2)(m-4)}{(m-3)(m-5)\sin.^{m-5} x}+\ldots\right\}+\frac{(m-2)(m-4)(m-6)\ldots}{(m-1)(m-3)(m-5)\ldots}\left\{\begin{array}{l}\log.\ \tan. \tfrac{1}{2} x \\ \text{or} \\ x,\end{array}\right.$$

the upper factor, log. tan. $\tfrac{1}{2}x$, to be used when m is odd and the lower, x, when m is even.

The terms within the brackets are to be continued as far as the exponents continue greater than 0, but no farther; and the exponent at which we stop will be the last factor in the denominator of the term without the brackets.

Let now $\frac{\pi}{2} - x$ be substituted for x, in this formula, and let m be changed into n, then, because $d\left(\frac{\pi}{2} - x\right) = - dx$, we shall have the following formula:

VI.

$$\int \frac{dx}{\cos.^n} =$$

$$\frac{\sin. x}{n-1} \left\{ \frac{1}{\cos.^{n-1} x} + \frac{n-2}{(n-3)\cos.^{n-3} x} + \frac{(n-2)(n-4)}{(n-3)(n-5)\cos.^{n-5} x} + \ldots \right\} + \frac{(n-2)(n-4)(-)\ldots}{(n-1)(3.5)(-)\ldots} \left\{ \begin{array}{l} \log. \tan. \frac{1}{2} (\frac{\pi}{2} + x), \\ \text{or} \\ x. \end{array} \right.$$

the upper factor belonging to the case of m, odd, and the lower to the case of m even. This upper factor is obtained in manner similar to that by which the corresponding factor in the preceding formula was found, thus the final integral being here $\int \frac{dx}{\cos. x}$ we have

$$\int \frac{dx}{\cos. x} = \int \frac{\cos. x\, dx}{1-\sin.^2 x} = \int \frac{d \sin. x}{1-\sin.^2 x} = \log. \left(\frac{1+\sin. x}{1-\sin. x}\right)^{\frac{1}{2}} = \log. \tan. \frac{1}{2} (\frac{\pi}{2} + x)^*.$$

---

* Dr. Gregory's Trigonometry, page 48.

EXAMPLES.

(38.)   1. To integrate

$$\frac{\cos.^7 x\, dx}{\sin. x}.$$

Applying here the formula II. we have

$$\int \frac{\cos.^7 x\, dx}{\sin. x} = \frac{1}{6} \left\{ \cos.^6 x + \frac{6}{4} \cos.^4 x + \frac{6 \cdot 4}{4 \cdot 2} \cos.^2 x \right\} + \int \frac{\cos. x\, dx}{\sin. x}$$

$$= \frac{1}{6} \cos.^6 x + \frac{1}{4} \cos.^4 x + \frac{1}{2} \cos.^2 x + \log. \sin. x + C.$$

2. To integrate

$$\frac{\sin.^4 x\, dx}{\cos.^5 x}.$$

Applying first the formula I. we have

$$\int \frac{\sin.^4 x\, dx}{\cos.^5 x} = \frac{1}{\cos.^4 x} \left\{ \sin.^3 x - \sin. x \right\} + \int \frac{dx}{\cos.^5 x},$$

and applying the formula VI. to this last integral we have

$$\int \frac{dx}{\cos.^5 x} = \frac{\sin. x}{4} \left\{ \frac{1}{\cos.^4 x} + \frac{3}{2 \cos.^2 x} \right\} + \frac{3}{4 \cdot 2} \log. \tan. \tfrac{1}{2} \left(\frac{\pi}{2} + x \right) + C,$$

so that

$$\int \frac{\sin.^4 x\, dx}{\cos.^5 x} = \frac{1}{\cos.^4 x} \left\{ \sin.^3 x - \frac{5}{4} \sin. x \right\} -$$

$$\frac{3}{4 \cdot 2} \left\{ \frac{\sin. x}{\cos.^2 x} + \log. \tan. \tfrac{1}{2} \left(\frac{\pi}{2} + x \right) \right\} + C.$$

3. To integrate

$$\frac{\cos.^6 x\, dx}{\sin. x}$$

$$\int \frac{\cos.^6 x\, dx}{\sin. x} = \frac{1}{5} \cos.^5 x + \frac{1}{3} \cos.^3 x + \cos. x + \log. \tan. \frac{1}{2} x + C.$$

4. To integrate

$$\frac{dx}{\cos.^4 x}$$

$$\int \frac{dx}{\cos.^4 x} = \frac{\sin. x}{3 \cos.^3 x} + \frac{2}{3} \tan. x + C.$$

5. To integrate

$$\frac{\sin.^3 x \, dx}{\cos.^2 x}$$

$$\int \frac{\sin.^3 x \, dx}{\cos.^2 x} = \cos. x + \sec. x + C.$$

It must be remarked that, in the form just considered, the process fails when $m = -n$, because then formulas I. and II. become inapplicable, but they are not needed in this case, for since

$$\frac{\sin.^n x}{\cos.^n x} = \tan.^n x, \quad \frac{\cos.^n x}{\sin.^n x} = \frac{1}{\tan.^n x},$$

if we put

$$\tan. x = y \; \therefore \; dx = \frac{dy}{1 + y^2}$$

we have

$$\int \frac{\sin.^n x}{\cos.^n x} \, dx = \int \frac{y^n}{1 + y^2} \, dy, \; \int \frac{\cos.^n x}{\sin.^n x} \, dx = \int \frac{1}{y^n (1 + y^2)} \, dy,$$

and therefore the proposed forms become reducible to rational fractions.

(39.) Before dismissing the preceding forms we ought to remark, that in those particular cases in which the exponents m and n are positive whole numbers, the integration may be effected without introducing any powers of the trigonometrical lines, the sines and cosines of multiple arcs occurring instead, and these are more easily calculated than the powers.

This form of the integral requires the development of $\sin.^m x$, $\cos.^n x$ in a finite series involving only the sines and cosines of the multiples of x, and which development is always possible when m and n are positive integers. It gives (*see Note B.*)

$$\sin.^2 x = -\frac{1}{2} \cos. 2x + \frac{1}{2}$$

$$\sin.^3 x = -\frac{1}{4} \sin. 3x + \frac{3}{4} \sin. x$$

$$\sin.^4 x = \frac{1}{8} \cos. 4 x - \frac{1}{2} \cos. 2 x + \frac{3}{8}$$

&c.   &c.

$$\cos.^2 x = \frac{1}{2} \cos. 2 x + \frac{1}{2}$$

$$\cos.^3 x = \frac{1}{4} \cos. 3 x + \frac{3}{4} \cos. x$$

$$\cos.^4 x = \frac{1}{8} \cos. 4 x + \frac{1}{2} \cos. 2 x + \frac{3}{8}$$

&c.   &c.

If, therefore, we multiply by dx and integrate we get

$$\int \sin.^2 x \, dx = - \frac{1}{2} \int \cos. 2 x \, dx + \frac{1}{2} x$$

$$= - \frac{1}{4} \sin. 2 x + \frac{1}{2} x + C$$

$$\int \sin.^3 x \, dx = - \frac{1}{4} \int \sin. 3 \, dx \, x + \frac{3}{4} \int \sin. x \, dx$$

$$= \frac{1}{12} \cos. 3 x + \frac{3}{4} \cos. x + C$$

$$\int \sin.^4 x \, dx = \frac{1}{8} \int \cos. 4 x \, dx - \frac{1}{2} \int \cos. 2 x \, dx + \frac{3}{8} x$$

$$= \frac{1}{32} \sin. 4 x + \frac{1}{4} \sin. 2 x + \frac{3}{8} x + C$$

&c.   &c.

$$\int \cos.^2 x \, dx = \frac{1}{2} \int \cos. 2 x \, dx + \frac{1}{2} x$$

$$= \frac{1}{4} \sin. 2 x + \frac{1}{2} x + C$$

$$\cos. \qquad \frac{1}{4} \int \cos. 3 x \, dx + \frac{3}{4} \int \cos. x \, dx$$

$$= \frac{1}{12} \sin. 3\, x + \frac{3}{4} \sin. x + C$$

$$\int \cos.^4 x\, dx = \frac{1}{8} \int \cos. 4\, x\, dx + \frac{1}{2} \int \cos. 2\, x\, dx + \frac{3}{8}\, x$$

$$= \frac{1}{32} \sin. 4\, x + \frac{1}{4} \sin. 2\, x + \frac{3}{8}\, x + C,$$

&c. &c.

By substituting in the expression for $\int \sin.^n x \cos.^m x\, dx$, for the powers of the sines and cosines the foregoing values, this integral also will be expressed without powers.

(40.) There still remains for us

To integrate the form

$$\frac{dx}{\sin.^m x \cos.^n x}.$$

Since this is the same as

$$\frac{\sin.^2 x\, dx + \cos.^2 x\, dx}{\sin.^m x \cos.^n x}$$

we have this decomposition, viz.

$$\int \frac{dx}{\sin.^m x \cos.^n x} = \int \frac{dx}{\sin.^{m-2} x \cos.^n x} + \int \frac{dx}{\sin.^m x \cos.^{n-2} x},$$

and, by decomposing in this way the successive component integrals each into two, we shall finally arrive at forms already integrated.

EXAMPLES.

(41.) 1. To integrate

$$\frac{dx}{\sin.^3 x \cos.^2 x}$$

$$\int \frac{dx}{\sin.^3 x \cos.^2 x} = \int \frac{dx}{\sin. x \cos.^2 x} + \int \frac{dx}{\sin.^3 x}$$

$$\int \frac{dx}{\sin. x \sin.^2 x} = \int \frac{\sin. x\, dx}{\cos.} + \int \frac{dx}{\cos. x} = \frac{1}{\cos. x} + \log. \tan. \frac{1}{2}\, x,$$

also (37)

$$\int \frac{dx}{\sin.^3 x} = -\frac{\cos. x}{2 \sin.^2 x} + \frac{1}{2} \log. \tan. \frac{1}{2} x,$$

consequently

$$\int \frac{dx}{\sin.^3 x \cos.^2 x} = \frac{1}{\cos. x} - \frac{\cos. x}{2 \sin.^2 x} + \frac{3}{2} \log. \tan. \frac{1}{2} x + C.$$

2. To integrate

$$\frac{dx}{\sin.^4 x \cos. x}$$

$$\int \frac{dx}{\sin.^4 x \cos. x} = \int \frac{dx}{\sin.^2 x \cos. x} + \int \frac{\cos. x dx}{\sin.^4 x}$$

$$\int \frac{dx}{\sin.^2 \cos. x} = \int \frac{dx}{\cos. x} + \int \frac{\cos. x dx}{\sin.^2 x}$$

$$= \log. \tan. \frac{1}{2} \left(\frac{1}{2} \pi + x\right) - \frac{1}{\sin. x}$$

also

$$\int \frac{\cos. x dx}{\sin.^4 x} = -\frac{1}{3 \sin.^3 x}$$

$$\therefore \int \frac{dx}{\sin.^4 x \cos. x} = \log. \tan. \frac{1}{2} \left(\frac{\pi}{2} + x\right) - \frac{1}{\sin. x} - \frac{1}{3 \sin.^3 x} + C.$$

3. To integrate

$$\frac{dx}{\sin.^n x \cos.^n}$$

This, since $\sin. x \cos. x = \sin. 2x$, is the same as

$$\frac{dx}{\sin.^n 2x} = \frac{2 dy}{\sin.^n y} \quad \therefore \int \frac{dx}{\sin.^n x \cos.^n x} = 2 \int \frac{dy}{\sin.^n y},$$

which form has already been integrated at (37).

4. To integrate

$$\frac{dx}{\sin.^2 x \cos.^2 x}$$

$$\int \frac{dx}{\sin.^2 x \cos.^2 x} = -2 \cot. 2x + C.$$

5. To integrate

$$\frac{dx}{\sin.^2 x \cos.^4 x}$$

$$\int \frac{dx}{\sin.^2 x \cos.^4 x} = \frac{\sin. x}{3 \cos.^3 x} + \frac{2}{3} \tan. x - 2 \cot. 2x + C.$$

(42.) From what has now been shewn it appears that the general differential expression

$$\sin.^m x \cos.^n x\, dx$$

may always be integrated when m and n are whole numbers, whether positive or negative. It is also completely integrable under other conditions and by the same formulas, as is easily seen by transforming it into an algebraic binomial differential, which we may always do. For if we put sin. $x = y$ we shall have cos. $x = \sqrt{1 - y^2}$, and the known expression for the differential of an arc x, in terms of its sine y, is (*Differential Calculus*, p. 20,)

$$dx = \frac{dy}{\sqrt{1 - y^2}};$$

hence, by substitution, the proposed differential takes the algebraic form

$$y^m (1 - y^2)^{\frac{n-1}{2}},$$

which we know may always be integrated when either $\dfrac{m+1}{2}$, or $\dfrac{m+n}{2}$ are whole numbers.

As in the preceding general formulas the exponents m or n are continually diminished by 2: this condition of integrability must necessarily subsist for the final or reduced integral.

The reduction of trigonometrical into algebraical functions is often advantageously adopted to facilitate the integration of such functions in cases which the preceding general formulas do not comprehend. But we shall not go into these cases here, as we propose to annex to the present section a supplementary chapter, exhibiting a specimen of those particular processes and transformations which are most frequently found to succeed when the integration is not to be effected by general rules.

(43.) There remains to be considered one more general trigonometrical form to which integration by parts as successfully applies as in the preceding cases.

To integrate the forms

$$m^{ax} \sin.^n x \, dx, \; m^{ax} \cos.^n x \, dx.$$

Taking the first of these forms, which may obviously be written

$$- m^{ax} \sin.^{n-1} x \, d \cos. x,$$

and putting in the formula for integration by parts

$$v = \cos. x, \; u = - m^{ax} \sin.^{n-1} x$$

$$\therefore du = - a.\log. m \cdot m^{ax} \sin.^{n-1} x \, dx + (n-1) \, m^{ax} \sin.^{n-2} \cos. x . dx$$

we have

$$- \int m^{ax} \sin.^{n-1} x \, d \cos. x = - m^{ax} \sin.^{n-1} x \cos. x \, +$$

$$a \log. m \int m^{ax} \sin.^{n-1} x \cos. x \, dx \, +$$

$$(n-1) \int m^{ax} \sin.^{n-2} x \cos.^2 x \, dx \; \ldots \; (1).$$

The first integral on the right is the same as

$$\frac{1}{n} \int m^{ax} \, d \sin.^n x = \frac{m^{ax} \sin.^n x}{n} - \frac{a \log. m}{n} \int m^{ax} \sin.^n x \, dx;$$

hence, by substitution, the equation (1) becomes

$$\int m^{ax} \sin.^n x \, dx = - m^{ax} \sin.^{n-1} x \cos. x + \frac{a \log. m \cdot m^{ax} \sin.^n x}{n} -$$

$$\frac{(a \log. m)^2}{n} \int m^{ax} \sin.^n x \, dx + (n-1) \int m^{ax} \sin.^{n-2} x \, (1 - \sin.^2 x) \, dx,$$

in the second member of which there are two integrals like that in the first member; therefore, by transposing these we have

$$\frac{(a \log. m)^2 + n^2}{n} \int m^{ax} \sin.^n x \, dx = - m^{ax} \sin.^{n-1} x \cos. x \, +$$

$$\frac{a \log. m \cdot m^{ax} \sin.^n x}{n} + (n-1) \int m^{ax} \sin.^{n-2} x \, dx,$$

consequently

$$\int m^{ax} \sin.^n x \, dx = \frac{m^{ax} \sin.^{n-1} x}{(a \log. m)^2 + n^2} \{ a \log. m \cdot \sin. x - n \cos. x \} \, +$$

$$\frac{n(n-1)}{(a \log. m)^2 + n^2} \int m^{ax} \sin.^{n-2} x \, dx \; \ldots \; (2).$$

By this formula therefore the proposed integral is reduced to another of the same form, but in which the exponent of sin. x is less by 2, so that if n be an even positive integer, we shall, by the successive application of (2), finally reduce the integration to

$$\int m^{ax}\, dx = \frac{1}{am}\, m^{ax} + C.$$

If n is an odd positive integer we shall arrive at the integral

$$\int m^{ax} \sin. x\, dx,$$

which is itself immediately given by the formula (2), since the factor $n-1$ then vanishes; therefore

$$\int m^{ax} \sin. x\, dx = \frac{m^{ax}}{(a \log. m)^2 + 1} \{a \log. m \cdot \sin. x - \cos. x\} + C.$$

By applying the same process to the other general form we shall obtain the formula

$$\int m^{ax} \cos.^n x\, dx = \frac{m^{ax} \cos.^{n-1} x}{(a \log. m)^2 + n^2} \{a \log. m \cdot \cos. x + n \sin. x\} +$$

$$\frac{n(n-1)}{(a \log. m)^2 + n^2} \int m^{ax} \cos.^{n-2} x\, dx.$$

EXAMPLES.

(44.) 1. To integrate

$$e^x \sin.^2 x\, dx$$

$$\int e^x \sin.^2 x\, dx = \frac{e^x \sin. x}{1 + 2^2} \{\sin. x - 2 \cos. x\} + \frac{2}{1 + 2^2} \int e^x\, dx$$

$$= \frac{e^x \sin. x}{5} \{\sin. x - 2 \cos. x\} + \frac{2}{5} e^x + C.$$

2. To integrate

$$e^{ax} \cos. x\, dx$$

$$\int e^{ax} \cos. x\, dx = \frac{e^{ax}}{a^2 + 1} \{a \cos. x + \sin. x\} + C.$$

3. To integrate

$$e^x \sin.^3 x \, dx$$

$$\int e^x \sin.^3 x \, dx = \frac{e^x}{10} \left\{ \sin.^3 x + 3 \cos.^3 x + 3 \sin. x - 6 \cos. x \right\} + C.$$

(45.) We shall terminate the present chapter by shewing how

To integrate the forms

$$X \sin.^{-1} x \, dx, \; X \cos.^{-1} x \, dx, \; \&c.$$

in which X is an algebraic function of x.

By applying to the first of these expressions the process of integration by parts, $\sin.^{-1} x$ being put for u and $\int X dx$ for v, we have, since

$$d \sin.^{-1} x = \frac{dx}{\sqrt{1 - x^2}},$$

the following formula, viz.

$$\int X \sin.^{-1} x \, dx = \sin.^{-1} x \int X dx - \int \frac{\int X dx}{\sqrt{1 - x^2}} \, dx,$$

which, since X is here supposed to be an algebraic function, reduces the proposed integration to that of the algebraic functions

$$\int X dx = X_1 \text{ and } \int \frac{X_,}{\sqrt{1 - x^2}} \, dx.$$

By applying the same process to the second expression, the integration of this also will be obviously reduced to that of algebraic forms; and such would always be the case if, in the above expression, $\tan.^{-1} x$, $\sec.^{-1} x$, &c. were put for $\sin.^{-1} x$, because the differentials of all these are algebraic functions.

(46.) We shall apply the above formula to one or two examples.

EXAMPLES.

1. To integrate

$$\frac{x^3 \, dx}{\sqrt{1 - x^2}} \sin.^{-1} x.$$

We have first to integrate

$$X dx = \frac{x^3\, dx}{\sqrt{1 - x^2}}.$$

By referring to ex. 3, p. 44, we find

$$\int \frac{x^3\, dx}{\sqrt{1 - x^2}} = -\left(\frac{1}{3}\, x^2 + \frac{1 \cdot 2}{1 \cdot 3}\right) \sqrt{1 - x^2};$$

also

$$\int \left(\frac{1}{3}\, x^2 + \frac{1 \cdot 2}{1 \cdot 3}\right) dx = \frac{x^2}{9} + \frac{2x}{3};$$

hence, by·substitution, the above formula becomes in the present case

$$\int \frac{x^3\, dx}{\sqrt{1 - x^2}} \sin.^{-1} x = -\left(\frac{1}{3}\, x^2 + \frac{2}{3}\right) \sqrt{1 - x^2} \cdot \sin.^{-1} x +$$

$$\frac{x^2}{9} + \frac{2x}{3} + C.$$

2. To integrate

$$x^2 \sin.^{-1} x\, dx$$

$$\int X dx = \int x^2\, dx = \frac{x^3}{3},$$

also

$$\int \frac{x^3\, dx}{3 \sqrt{1 - x^2}} = -\frac{1}{3}\left(\frac{1}{3}\, x^2 + \frac{2}{3}\right) \sqrt{1 - x^2};$$

hence

$$\int x^2 \sin.^{-1} x\, dx = \frac{x^3}{3} \sin.^{-1} x - \frac{1}{3}\left(\frac{1}{3}\, x^2 + \frac{2}{3}\right) \sqrt{1 - x^2} + C.$$

3·. To integrate

$$\frac{x^4\, dx}{\sqrt{1 - x^2}} \sin.^{-1} x$$

$$\int \frac{x^4\, dx}{\sqrt{1 - x^2}} \sin.^{-1} x = -\left\{\left(\frac{1}{4}\, x^3 + \frac{3}{2 \cdot 4}\, x\right) \sqrt{1 - x^2} - \frac{3}{16} \sin.^{-1} x\right\} +$$

$$\frac{1}{16}\, x^4 + \frac{3}{16}\, x^2 + C.$$

4. To integrate

$$x^4 \tan.^{-1} x\, dx$$

$$\int x^4 \tan.^{-1} x \, dx = \frac{x^5}{5} \tan.^{-1} x - \frac{1}{5} \left\{ \frac{x^4}{4} - \frac{x^2}{2} + \log. (1 + x^2) \right\} + C.$$

5. To integrate

$$x^m \sin.^{-1} x \, dx \text{ and } x^m \tan.^{-1} x \, dx$$

$$\int x^m \sin.^{-1} x \, dx = \frac{x^{m+1}}{m+1} \sin.^{-1} x - \frac{1}{m+1} \int \frac{x^{m+1} \, dx}{\sqrt{1-x^2}}$$

$$\int x^m \tan.^{-1} x \, dx = \frac{x^{m+1}}{m+1} \tan.^{-1} x - \frac{1}{m+1} \int \frac{x^{m+1} \, dx}{1+x^2} \; .$$

CHAPTER VI.

ON INTEGRATION BY SERIES, AND ON SUCCESSIVE INTEGRATION.

(47.) As our object hitherto has been to obtain general rules and formulas for the integration of a differential expression, we have confined our attention to the principal of those forms which are completely integrable. These, however, are very few, in comparison to those forms which the calculus in its present state furnishes no means of integrating in finite terms. We now come to consider this latter class of differentials, and to shew that the integral of any differential whatever may always be expressed by means of series.

Integration by Series.

(48.) Let X represent any function of x whatever, and put

$$\int X dx = Fx \; . \; . \; . \; . \; (1),$$

then, by Taylor's theorem,

$$F(x-h) = \int X dx - Xh + \frac{dX}{dx} \cdot \frac{h^2}{1 \cdot 2} - \frac{d^2X}{dx^2} \cdot \frac{h^3}{1 \cdot 2 \cdot 3} + \&c.$$

or, substituting x for the indeterminate h,

$$F(x - x) = [Fx] = \int X dx - X x + \frac{dX}{dx} \cdot \frac{x^2}{1 \cdot 2} - \frac{d^2X}{dx^2} \cdot \frac{x^3}{1 \cdot 2 \cdot 3} + \&c.$$

where $[Fx]$ is what Fx becomes when $x = 0$. Hence, by transposition, we have

$$\int X dx = X x - \frac{dX}{dx} \cdot \frac{x^2}{1 \cdot 2} + \frac{d^2x}{dx^2} \cdot \frac{x^3}{1 \cdot 2 \cdot 3} - \&c. + [\int X dx]...(2),$$

which is the series of *John Bernouilli.*

From this general expression for the integral of $X dx$ it appears that the integration of every differential expression, containing one variable, may always be obtained, although not always in finite terms. The quantity $[\int X dx]$ is obviously the arbitrary constant, being what the complete integral becomes when $x = 0$, that is, $[\int X dx] = C$.

We have given this series of Bernouilli more with the view of shewing the possibility of obtaining in every case an expression for the integral, than for the sake of the utility of this expression in computing the actual value of the integral in particular cases. For such purpose, it is obviously necessary that the series converge, which requires that it proceed according either to the ascending or the descending powers of the variable, which that above will rarely do, seeing that the several differential coefficients are functions of x.

(49.) If, instead of the theorem of Taylor, we apply that of Maclaurin to the function (1), we shall have

$$\int X dx = [\int X dx] + [X] x + [\frac{dX}{dx}] \frac{x^2}{1 \cdot 2} + [\frac{d^2X}{dx^2}] \frac{x^3}{1 \cdot 2 \cdot 3} + \&c. \ . \ . \ . \ (3),$$

which is a series much more useful for the purpose in question than that just given. This, however, fails to be applicable when $x = 0$ renders X, or $\frac{dX}{dx}$, or $\frac{d^2X}{dx^2}$, &c. also 0. The term $[\int X dx]$ is here, as before, the constant C, which completes the integral.

(50.) Another mode of obtaining the developed integral $\int X dx$, and the one most frequently employed, is to develope X by the processes of algebra into a series of terms, such that, being multiplied by dx, each may be integrable separately; then these series of integrals will necessarily be the development of $\int X dx$. In this way, we may readily derive the formula (3), above, belonging to those cases where X may be developed according to the increasing positive and whole powers of x. For such development of x, by whatever process obtained, must of course agree with that furnished by Maclaurin's theorem, that is

$$X = [X] + [\frac{dX}{dx}] x + [\frac{d^2X}{dx^2}] \frac{x^2}{1 \cdot 2} + \&c.$$

Now there are two ways equally obvious, in which we may render the right hand member of this equation identical to the right hand member of (3); we may, as above noticed, multiply by dx, and integrate each term, annexing the arbitrary constant C, or we may multiply by x, and then divide each term by the number deno·ing its place in the series, still annexing the constant C.

(51.) This latter method leads us to remark, in passing, that the development of functions by Maclaurin's theorem may sometimes be considerably facilitated by developing one of the differential coefficients algebraically, instead of continuing to differentiate.

Had this means of avoiding the trouble of differentiating occurred to us at the time, we should certainly have adopted it in developing tan.$^{-1} x$, at page 35 of the Differential Calculus. For, by developing by common division, the first differential coefficient

$$\frac{dy}{dx} = \frac{1}{1 + x^2},$$

we get with the greatest ease the series

$$1 + 0x - x^2 + 0x^3 + x^4 + 0x^5 - x^6 + \&c.$$

and it merely remains now to multiply this series by x, and to divide each term by the number denoting its place, so that, putting tan.y for x, we have

$$y = \tan. y - \frac{1}{3} \tan.^3 y + \frac{1}{5} \tan.^5 y - \frac{1}{7} \tan.^7 y + \&c.$$

and in a similar manner may the developments of sin.$^{-1} x$, cos.$^{-1} x$, &c. be facilitated, as will be farther shewn in some of the following examples :

EXAMPLES.

(52.) 1. To determine the integral of

$$\frac{dx}{a+x}$$

in a series.

By division

$$\frac{1}{a+x}=\frac{1}{a}-\frac{x}{a^2}+\frac{x^2}{a^3}-\frac{x^3}{a^4}+\&c.$$

Multiplying by x, and dividing each term on the right by the number denoting its place, we have

$$\int\frac{dx}{a+x}=\frac{x}{a}-\frac{x^2}{2a^2}+\frac{x^3}{3a^3}-\frac{x^4}{4a^4}+\&c.+C.$$

We already know, however, that

$$\int\frac{dx}{a+x}=\log.(a+x)+C,$$

or, in other words, that $\dfrac{dx}{a+x}$ is the first differential coefficient of $\log.(a+n)$, so that, by means of this first coefficient only we easily get the development

$$\log.(a+x)=\log.a+\frac{x}{a}-\frac{x^2}{2a^3}+\frac{x^3}{3a^3}-\frac{x^4}{4a^4}+\&c.$$

the first term of the development, when it proceeds according to the positive integral powers of x, being always what the proposed function becomes when $x=0$.

2. To develope

$$\int\frac{dx}{\sqrt{1+x^2}}.$$

By the binomial theorem

$$\frac{1}{\sqrt{1+x^2}}=1-\frac{1}{2}x^2+\frac{1\cdot3}{2\cdot4}x^4-\frac{1\cdot3\cdot5}{2\cdot4\cdot6}x^6+\&c.$$

Multiplying by x, and dividing the several terms on the right by 1, 3, 5, &c. on account of the absent terms in x, x^3, &c. we have

$$\int \frac{dx}{\sqrt{1+x^2}} = x - \frac{1}{2\cdot 3}x^3 + \frac{1\cdot 3}{2\cdot 4\cdot 5}x^4 - \frac{1\cdot 3\cdot 5}{2\cdot 4\cdot 6\cdot 7}x^7 + \&c. + C.$$

But (p. 34)

$$\int \frac{dx}{\sqrt{1+x^2}} = \log.(x + \sqrt{1+x^2}) + C,$$

that is, the first differential coefficient of $\log.(x + \sqrt{1+x^2})$ is $\dfrac{dx}{\sqrt{1+x^2}}$ by means of which we obtain the development

$$\log.(x + \sqrt{1+x^2}) = x - \frac{1}{2\cdot 3}x^3 + \frac{1\cdot 3}{2\cdot 4\cdot 5}x^4 - \frac{1\cdot 3\cdot 5}{2\cdot 4\cdot 6\cdot 7}x^7 + \&c.$$

3. To integrate

$$\frac{dx}{\sqrt{1-x^2}}$$

by series.

By the binomial theorem

$$\frac{1}{\sqrt{1-x^2}} = 1 + \frac{1}{2}x^2 + \frac{1\cdot 3}{2\cdot 4}x^4 + \frac{1\cdot 3\cdot 5}{2\cdot 4\cdot 6} + \&c.$$

Multiplying by x and dividing the terms by 1, 3, 5, &c. severally we have

$$\int \frac{dx}{\sqrt{1-x^2}} = x + \frac{1}{2\cdot 3}x^3 + \frac{1\cdot 3}{2\cdot 4\cdot 5}x^5 + \frac{1\cdot 3\cdot 5}{2\cdot 4\cdot 6.7}x^7 + C.$$

But

$$\int \frac{dx}{\sqrt{1-x^2}} = \sin.^{-1}x$$

consequently

$$\sin.^{-1}x = x + \frac{1}{2\cdot 3}x^3 + \frac{1\cdot 3}{2\cdot 4\cdot 5}x^5 + \frac{1\cdot 3\cdot 5}{2\cdot 4\cdot 6\cdot 7}x^7.$$

4. To integrate

$$\frac{dx}{\sqrt{x^2-1}}$$

in a series of descending powers of x

By the binomial theorem

$$\frac{1}{\sqrt{x^2-1}} = \frac{1}{x\sqrt{1-\frac{1}{2x^2}}} = \frac{1}{x} + \frac{1}{2x^3} + \frac{1\cdot 3}{2\cdot 4x^5} +$$

$$\frac{1\cdot 3\cdot 5}{2\cdot 4\cdot 6x^7} + \&c.$$

Multiplying by dx, and integrating each term, we have

$$\int \frac{dx}{\sqrt{x^2-1}} = \log. x - \frac{1}{2\cdot 2x^2} - \frac{1\cdot 3}{2\cdot 4\cdot 4x^4} -$$

$$\frac{1\cdot 3\cdot 5}{2\cdot 4\cdot 6\cdot 6\, x^6} - \&c. + C,$$

but (18)

$$\int \frac{dx}{\sqrt{x^2-1}} = \log. (x + \sqrt{x^2-1}) + C;$$

hence, by this means we obtain the development

$$\log. (x + \sqrt{x^2-1}) = \text{First term} + \log. x - \frac{1}{2\cdot 2x^2} - \frac{1\cdot 3}{2\cdot 4\cdot 4x^4} -$$

$$\frac{1\cdot 3\cdot 5}{2\cdot 4\cdot 6\cdot 6x^6} - \&c.$$

As the series in this case is not according to the positive powers of x, it does not agree with Maclaurin's, and, therefore, the first term is not what the proposed becomes when $x = 0$; we may, however, easily discover what this term should be. We at once see that by putting $x = 1$ the first member of this equation becomes $\log. 1 = 0$, and the second becomes

$$\text{First term} - \frac{1}{2\cdot 2} - \frac{1\cdot 3}{2\cdot 4\cdot 4} - \frac{1\cdot 3\cdot 5}{2\cdot 4\cdot 6\cdot 6} - \&c.$$

which must, of course, be also 0, consequently

$$\text{First term} = \frac{1}{2\cdot 2} + \frac{1\cdot 3}{2\cdot 4\cdot 4} + \frac{1\cdot 3\cdot 5}{2\cdot 4\cdot 6\cdot 6} - \&c.$$

so that we thus have the complete development of $\log. (x + \sqrt{x^2-1}.)$

5. To integrate

$$\frac{dx}{x^2 + 1}$$

in a series of descending powers of x.

By division

$$\frac{1}{x^2 + 1} = \frac{1}{x^2 \left(1 + \dfrac{1}{x^2}\right)} = \frac{1}{x^2} - \frac{1}{x^4} + \frac{1}{x^6} - \frac{1}{x^8} + \&c.$$

Multiplying by dx and integrating we have

$$\int \frac{dx}{x^2 + 1} = -\frac{1}{x} + \frac{1}{3x^3} - \frac{1}{5x^5} + \frac{1}{7x^7} - \&c. + C;$$

hence

$$\tan.^{-1} x = \text{First term} - \frac{1}{x} + \frac{1}{3x^3} - \frac{1}{5x^3} + \frac{1}{7x^7} - \&c.$$

To determine the first term we may remark that when $x = \infty$, $\tan.^{-1} x = \dfrac{\pi}{2}$, so that the development above then becomes

$$\frac{\pi}{2} = \text{First term},$$

the first term is therefore thus determined.

The preceding examples will serve to shew that functions may sometimes be readily developed, by first developing the differential, and then integrating each term separately.

6. To determine the integral of

$$\frac{\sqrt{1 - e^2 x^2}}{\sqrt{1 - x^2}} \, dx.$$

This differential cannot be integrated by any of the formulas in the preceding chapter: but, since it may be written

$$\frac{dx}{\sqrt{1 - x^2}} \cdot \sqrt{1 - e^2 x^2};$$

it is obvious that if $\sqrt{1 - e^2 x^2}$ be developed in a series of ascending powers of x, the development of the proposed will be a series of terms all of the form

$$\frac{x^m \, dx}{\sqrt{1 - x^2}},$$

and are therefore all integrable (25).

By the binomial theorem

$$\sqrt{1-e^2\,x^2}=1-\frac{1}{2}\,e^2\,x^2+\frac{1}{2\cdot4}\,e^4\,x^4-\frac{1\cdot3}{2\cdot4\cdot6}\,e^6\,x^6+\&c.$$

and multiplying by

$$\frac{dx}{\sqrt{1-x^2}},$$

and then, integrating each term, there results

$$\int\frac{\sqrt{1-e^2\,x^2}}{\sqrt{1-x^2}}\,dx=$$

$\sin.^{-1}x$

$$+\frac{1}{2}\,e^2\,\{\frac{1}{2}\,x\,\sqrt{1-x^2}-\frac{1}{2}\,\sin.^{-1}x\}$$

$$+\frac{1}{2\cdot4}\,e^4\,\{(\frac{1}{4}\,x^3+\frac{1\cdot3}{2\cdot4}x)\,\sqrt{1-x^2}-\frac{1\cdot3}{2\cdot4}\,\sin.^{-1}x\}$$

$$+\frac{2\cdot4\cdot6}{2\cdot4\cdot6}\,e^6\,\{(\frac{1}{6}\,x^5+\frac{1\cdot5}{4\cdot6}\,x^3+\frac{1\cdot3\cdot5}{2\cdot4\cdot6}\,x)\sqrt{1-x^2}-\frac{1\cdot3\cdot5}{2\cdot4\cdot6}\,\sin^{-1}x\}.$$

$$+\qquad\&c.\qquad\qquad+C.$$

7. To determine the integral of

$$\frac{a^x\,dx}{1-x}.$$

By division

$$\frac{1}{1-x}=1+x+x^2+x^3+\&c.\;\;\ldots\;\;(1),$$

also (*Diff. Calc.* p. 29,)

$$a^x=1+x\log.\,a+\frac{1}{2}\,x^2\log.^2a+\frac{1}{2\cdot3}\,x^3\log.^3a+\&c.\;\;\ldots\;\;(2).$$

Multiplying (1) and (2) together, we easily get

$$\frac{a^x}{1-x}=1+(1+\log.\,a)\,x+(1+\log.\,a+\frac{\log.^2a}{2})\,x^2+(1+\log.\,a+$$

$$\frac{\log.^2 a}{2} + \frac{\log.^3 a}{2 \cdot 3}) \, x^3 + \&c.$$

consequently

$$\int \frac{a^x \, dx}{1-x} = x + (1 + \log. a) \frac{x^2}{2} + (1 + \log. a + \frac{\log.^2 a}{2}) \frac{x^3}{3} +$$

$$(1 + \log. a + \frac{\log.^2 a}{2} + \frac{\log.^3 a}{2 \cdot 3}) \frac{x^4}{4} + \&c. + C.$$

8. To integrate

$$\frac{dx}{\sqrt{2x - x^2}}$$

by series.

$$\int \frac{dx}{\sqrt{2x - x^2}} = \sqrt{\frac{x}{2}} \{2 + \frac{x}{2 \cdot 3} + \frac{3x^2}{2 \cdot 4 \cdot 5 \cdot 2} +$$

$$\frac{3 \cdot 5 x^3}{2 \cdot 4 \cdot 6 \cdot 7 \cdot 4} + \&c.\} + C$$

$$= \text{versin.}^{-1} x + C.$$

9. To integrate

$$\frac{dx}{\sqrt{2x + x^2}}$$

by series.

$$\int \frac{dx}{\sqrt{2x + x^2}} = \sqrt{2x} \{1 - \frac{1}{2 \cdot 3} \cdot \frac{x}{2} + \frac{1 \cdot 3}{2 \cdot 4 \cdot 5} \cdot \frac{x^2}{4} -$$

$$\frac{1 \cdot 3 \cdot 5}{2 \cdot 4 \cdot 6 \cdot 7} \cdot \frac{x^3}{8} + \&c.\} + C$$

$$= \log. C (x + 1 + \sqrt{2x + x^2}) \; (\text{ex. } 12, \text{ p. } 34.)$$

10. To determine

$$\int \frac{dy}{\log. y} \, .$$

$$\int \frac{dy}{\log. y} = (\log.)^2 y + \log. y + \frac{1}{1 \cdot 2} \; \frac{\log.^2 y}{2} +$$

$$\frac{1}{1 \cdot 2 \cdot 3} \cdot \frac{\log.^3 y}{3} + \&c. + C.$$

11. To integrate

$$(a + bx^n)^{\frac{p}{q}} x^{m-1} dx,$$

when it does not satisfy either of the conditions of integrability,

$$\int (a + bx^n)^{\frac{p}{q}} x^{m-1} dx =$$

$$a^{\frac{p}{q}} \left\{ \frac{x^m}{m} + \frac{pb}{qa} \frac{x^{m+n}}{m+n} + \frac{p(p-q) b^2}{1 \cdot 2q^2 a^2} \frac{x^{m+2n}}{m+2n} + \&c. \right\} + C.$$

12. To integrate the same form in a series of descending powers of x.

Putting the differential under the form

$$(b + \frac{a}{x^n})^{\frac{p}{q}} x^{\frac{m+np}{q}-1} dx,$$

and developing the first factor by the binomial theorem, we find by integrating each term

$$\int (a + bx^n)^{\frac{p}{q}} x^{m-1} dx =$$

$$qb^{\frac{p}{q}} \left\{ \frac{x^{\frac{mq+np}{q}}}{mq+np} + \frac{pa}{qb} \frac{x^{mq+n(p-q)}}{mq+n(p-q)} + \frac{p(p-q)a^2}{1 \cdot 2q^2 b^2} \frac{x^{mq+n(p-2q)}}{mq+n(p-2q)} \right.$$

$$\left. + \&c. \right\} + C.$$

Successive Integration.

(53.) In all the foregoing examples, the first differential coefficient is given to determine the primitive function from which it has been derived; when, however, it is not the first, but the nth differential coefficient which is given, then by a first integration, we shall arrive at the preceding or $n - 1$th differential coefficient; by a second integration we get the $n - 2$th coefficient; and, by thus continuing the integration, we at length arrive at the original function. As each integration introduces a constant, it follows that the complete primitive ought to contain as many arbitrary constants as it has required integraions to obtain it.

Let y represent the primitive function, x being the variable, and put

$$\frac{d^n y}{d x^n} = X \therefore d\left(\frac{d^{n-1} y}{d x^{n-1}}\right) = X dx;$$

hence, by integrating, we have

$$\frac{d^{n-1} y}{d x^{n-1}} = \int X dx = X_1 + C_1.$$

Again, from this last equation we get

$$d\left(\frac{d^{n-2} y}{d x^{n-2}}\right) = X_1 dx + C_1 dx,$$

and, by integrating,

$$\frac{d^{n-2} y}{d x^{n-2}} = \int X_1 dx + \int C_1 dx = X_2 + C_1 x + C_2.$$

In like manner, from this we obtain

$$d\left(\frac{d^{n-3} y}{d x^{n-3}}\right) = X_2 dx + C_1 x dx + C_2 dx,$$

and integrating

$$\frac{d^{n-3} y}{d x^{n-3}} = \int X_2 dx + \int C_1 x dx + \int C_2 dx = X_3 + C_1 \frac{x^2}{2} + C_2 x + C_3;$$

and, continuing this process, we have, after n integrations,

$$\int^n X dx^n = X_n + C_1 \frac{x^{n-1}}{2 \cdot 3 \ldots (n-1)} + C_2 \frac{x^{n-2}}{2 \cdot 3 \ldots (n-2)} +$$

$$C_3 \frac{x^{n-3}}{2 \cdot 3 \ldots (n-3)}$$

$$+ \ldots + C_{n-2} \frac{x^2}{2} + C_{n-1} x + C_n.$$

The first term X_n of this series is the nth integral of $X dx^n$, without the arbitrary constants; the remaining part of the series is what ought to be annexed to every such integral in order to render it complete.

(54.) We may readily obtain the development of $\int^n X dx$ as follows:
By Maclaurin's theorem,

$$\int^n X dx^n = [\int^n X dx^n] + [\int^{n-1} dx^{n-1}] x + [\int^{n-2} X dx^{n-2}] \frac{x^2}{1 \cdot 2} +$$

$$\ldots\ldots + [\int X dx] \frac{x^{n-1}}{1 \cdot 2 \ldots (n-1)} + [X] \frac{x^n}{1 \cdot 2 \ldots . n} +$$

$$[\frac{dX}{dx}] \frac{x^{n+1}}{1 \cdot 2 \ldots (n+1)} + [\frac{d^2 X}{dx^2}] \frac{x^{n+2}}{1 \cdot 2 \ldots (n+2)} + \&c.$$

in which

$$[\int X dx] , [\int^2 X dx^2] \ldots\ldots [\int^n X dx^n]$$

are the constants $C_1, C_2 \ldots\ldots C_n$.

It appears from this formula that when $\int X dx$ is developable according to the increasing positive whole powers of x, so also is $\int^n X dx^n$, and that in such cases it is nearly as easy to determine by this formula the complete integral of $\int^n X dx^n$ as that of $\int X dx$; it will be necessary merely to develop X according to the increasing powers of x, as in the former parts of this chapter, and to substitute for x^0, x, x^2, x^3, &c. in that development, the quantities

$$\frac{x^n}{1 \cdot 2 \ldots . n} , \frac{x^{n+1}}{1 . 2 \ldots (n+1)} , \frac{x^{n+2}}{3 \cdot 4 \ldots (n+2)} , \frac{x^{n+3}}{4 \cdot 5 \ldots (n+3)} , \&c.$$

annexing the terms containing the arbitrary constants as above exhibited.

EXAMPLES.

(55.) 1. To determine

$$\int^4 \frac{dx^4}{\sqrt{1 + x^2}}.$$

By the binomial theorem

$$\frac{1}{\sqrt{1 + x^2}} = 1 - \frac{1}{2} x^2 + \frac{1 \cdot 3}{2 \cdot 4} x^4 - \frac{1 \cdot 3 \cdot 5}{2 \cdot 4 \cdot 6} x^6 + \&c.$$

and if in this series we substitute, agreeably to the above directions, instead of x^0, x^2, x^4, x^6, &c. the quantities

$$\frac{x^4}{1 \cdot 2 \cdot 3 \cdot 4} , \frac{x^6}{3 \cdot 4 \cdot 5 \cdot 6} , \frac{x^8}{5 \cdot 6 \cdot 7 \cdot 8} , \frac{x^{10}}{7 \cdot 8 \cdot 9 \cdot 10} , \&c.$$

we shall have

$$\int^4 \frac{dx^4}{\sqrt{1+x^2}} = \frac{x^4}{2\cdot3\cdot4} - \frac{x^6}{2\cdot3\cdot4\cdot5\cdot6} + \frac{1\cdot3x^8}{2\cdot4\cdot5\cdot6\cdot7\cdot8} -$$

$$\frac{1\cdot3\cdot5x^{10}}{2\cdot4\cdot6\cdot7\cdot8\cdot9\cdot10} + \&c.$$

$$+ C_1 \frac{x^3}{2\cdot3} + C_2 \frac{x^2}{2} + C_3 x + C_4.$$

2. To determine

$$\int^3 \sin. x \, dx^3.$$

By actually integrating, omitting the constants,

$$\int \sin. x \, dx = -\cos. x$$

$$\int^2 \sin. x \, dx^2 = -\int \cos. x \, dx = -\sin. x$$

$$\int^3 \sin. x \, dx^3 = -\int \sin. x \, dx = \cos. x$$

hence, completing the integral,

$$\int^3 \sin. x \, dx^3 = \cos. x + C_1 \frac{x^2}{2} + C_2 x + C_3.$$

3. Required the curve whose equation is

$$\frac{d^3y}{dx^3} = C_1 \text{ or } \frac{d^4y}{dx^4} = 0.$$

Here $X_1 \, X_2, \, X_3, \, X_4,$ (art. 51,) are each 0, therefore

$$y = C_1 \frac{x^3}{2\cdot3} + C_2 \frac{x^2}{2} + C_3 x + C_4;$$

hence the curve is a parabola of the third order, or else one of its varieties.

4. To determine

$$\int^2 \frac{a^2 - x^2}{(a^2 + a^2)^2} \, dx^2.$$

Decomposing the fraction, we find

$$-\frac{x^2 - a}{(x^2 + a^2)^2} = \frac{2x^2}{(x^2 + a^2)^2} - \frac{1}{x^2 + a^2},$$

multiplying by dx, and integrating, we have (13)

$$\int \frac{2x^2\,dx}{(x^2+a^2)^2} = -\frac{x}{x^2+a^2} + \int \frac{dx}{x^2+a^2}$$

and adding

$$-\int \frac{dx}{x^2+a^2}$$

$$\therefore \int \frac{a^2-x^2}{(a^2+x^2)^2}\,dx = \frac{x}{x^2+a^2} + C.$$

therefore

$$\int^2 \frac{a^2-x^2}{(x^2+a^2)^2}\,dx^2 = \int \frac{x\,dx}{x^2+a^2} + \int C\,dx$$

$$= \frac{1}{2}\log.\,(x^2+a^2) + Cx + C_1.$$

5. To determine

$$\int^4 \cos.\,x\,dx^4.$$

$$\int^4 \cos.\,x\,dx = \cos.x + C_1\frac{x^3}{2\cdot 3} + C_2\frac{x^2}{2} + C_3\,x + C_4.$$

6. To determine

$$\int^3 e^x\,dx^3.$$

$$\int^3 e^x\,dx^3 = e^x + C_1\frac{x^2}{2} + C_2\,x + C_3.$$

7. To determine

$$\int^3 \frac{dx^4}{\sqrt{1-x^2}}.$$

$$\int^4 \frac{dx^4}{\sqrt{1-x^2}} = \frac{x^4}{2\cdot 3\cdot 4} + \frac{x^6}{2\cdot 3\cdot 4\cdot 5\cdot 6} + \frac{1\cdot 3x^8}{2\cdot 4\cdot 5\cdot 6\cdot 7\cdot 8} +$$

$$\frac{1\cdot 3\cdot 5x^{10}}{2\cdot 4\cdot 6\cdot 7\cdot 8\cdot 9\cdot 10} + \&c.$$

$$+ C_1\frac{x^3}{2\cdot 3} + C_2\frac{x^2}{2} + C_3\,x + C_4.$$

CHAPTER VII.

ON INTEGRATION BETWEEN LIMITS, AND ON THE SUMMATION SERIES.

(56.) In the practical applications of the calculus, it is not the general, or, as it is usually called, the *indefinite*, integral that is ultimately required, because here the constant which completes the integral is indeterminate, whereas, in every particular inquiry this constant has a corresponding particular value, thus rendering the integral *definite*.

When the indefinite integral is found, it is easily rendered definite by the nature of the problem, which always fixes a limit or *origin* to the integral, that is, it is known to become 0 for some known value of the variable, and from this circumstance the proper value of the constant becomes determinable (3). The integral is, indeed, in most cases entirely limited by the nature of the problem, being comprised between two given values of the variable $x = a$, $x = b$; so that, by substituting these values successively in the general expression, and taking the dif ference of the results, the arbitrary constant becomes eliminated, and the remainder, which is entirely definite, is the value of the integral between the proposed limits. Thus the general or indefinite integral of $x^m dx$ is

$$\int x^m dx = \frac{x^{m+1}}{m+1} + C:$$

but, if this be required between the limits $x = a$, $x = b$, we shall have to take the difference between

$$\frac{a^{m+1}}{m+1} + C \text{ and } \frac{b^{m+1}}{m+1} + C,$$

and we express the result of this integration between the limits a and b thus:

$$\int_b^a x^m dx = \frac{a^{m+1} - b^{m+1}}{m+1},$$

the limit b, corresponding to the value which is subtracted, being placed below the other limit a

If b be the origin of the bove general integral, or the value of x for which it vanishes, then, since

$$0 = \frac{b^{m+1}}{m+1} + C,$$

$$\therefore C = -\frac{b^{m+1}}{m+1},$$

so that the definite integral is

$$\int_b^x x^m \, dx = \frac{x^{m+1} - b^{m+1}}{m+1} :$$

we shall give an example or two, in which integration between limits will be required.

EXAMPLES.

(57.)   1. It is an important question in mechanics to determine the time which a heavy body will require to fall through an arc of a vertical circle, that is to say, from a proposed point of departure to the lower extremity of the vertical diameter. If a represent the radius of the circle,h the height of the point of departure, and x any variable intermediate height from 0 to h, we are led, by the laws of motion, to an expression for the differential of the time containing the differential

$$\frac{dx}{\sqrt{(2ax - x^2)(h - x)}},$$

this therefore must be integrated, to obtain the time sought. It is not, however, integrable in finite terms, but by writing it thus

$$\frac{dx}{\sqrt{hx - x^2}} \left(1 - \frac{x}{2a}\right)^{-\frac{1}{2}},$$

we at once see that, if the second factor be developed in a series of ascending powers of x, the proposed differential will be reduced to a series of others, all of the form

$$\frac{x^m \, dx}{\sqrt{hx - x^2}},$$

which is an integrable form.

Therefore, developing by the binomial theorem, we have

$$\left(1 - \frac{x}{2a}\right)^{-\frac{1}{2}} = 1 + \frac{1}{2} \; \frac{x}{2a} + \frac{1 \cdot 3}{2 \cdot 4} \; \frac{x^2}{4a^2} + \frac{1 \cdot 3 \cdot 5}{2 \cdot 4 \cdot 6} \; \frac{x^3}{8a^3} + \&c.$$

and multiplying each term of this series by the other factor, we have this series of integrals

$$\int \frac{dx}{\sqrt{hx - x^2}} = \frac{2}{h} \text{ versin}^{-1} x \text{ to radius } \frac{h}{2} + C$$

$$\int \frac{x\,dx}{\sqrt{hx - x^2}} = -\sqrt{hx - x^2} + \frac{h}{2} \int \frac{dx}{\sqrt{hx - x^2}}$$

$$\int \frac{x^2\,dx}{\sqrt{hx - x^2}} = -\frac{x\sqrt{hx - x^2}}{2} + \frac{3h}{4} \int \frac{x\,dx}{\sqrt{hx - x^2}}$$

$$\int \frac{x^3\,dx}{\sqrt{hx - x^2}} = -\frac{x^2\sqrt{hx - x^2}}{3} + \frac{5h}{6} \int \frac{x^2\,dx}{\sqrt{hx - x^2}}$$

$$\&c. \qquad\qquad \&c.$$

By the question the limits of these integrals are $x = 0$, $x = h$; we have, therefore,

$$\int_0^h \frac{dx}{\sqrt{hx - x^2}} = \pi$$

$$\therefore \int_0^h \frac{x\,dx}{\sqrt{hx - x^2}} = \frac{h}{2} \pi$$

$$\int_0^h \frac{x^2\,dx}{\sqrt{hx - x^2}} = \frac{1 \cdot 3 h^2}{2 \cdot 4} \pi$$

$$\int_0^h \frac{x^3\,dx}{\sqrt{hx - x^2}} = \frac{1 \cdot 3 \cdot 5 h^3}{2 \cdot 4 \cdot 6} \pi$$

$$\&c.$$

and, consequently,

$$\int_0^h \frac{dx}{\sqrt{(2ah - x^2)(h - x)}} =$$

$$\left\{1 + \left(\frac{1}{2}\right)^2 \frac{h}{2a} + \left(\frac{1 \cdot 3}{2 \cdot 4}\right)^2 \frac{h^2}{2^2 a^2} + \left(\frac{1 \cdot 3 \cdot 5}{2 \cdot 4 \cdot 6}\right)^2 \frac{h^3}{2^2 a^3} + \&c.\right\} \pi.$$

    3. To determine

$$\int_0^1 \frac{x^m\,dx}{\sqrt{1 - x^2}}.$$

By taking the difference between the two values of each integral at pages 44 and 45, for the values $x = 0$ and $x = 1$, we have

$$\int_0^1 \frac{dx}{\sqrt{1-x^2}} = \frac{\pi}{2} \qquad\qquad \int_0^1 \frac{x\,dx}{\sqrt{1-x^2}} = 1$$

$$\int_0^1 \frac{x^2\,dx}{\sqrt{1-x^2}} = \frac{1}{2}\cdot\frac{\pi}{2} \qquad\qquad \int_0^1 \frac{x^3\,dx}{\sqrt{1-x^2}} = \frac{2}{3}$$

$$\int_0^1 \frac{x^4\,dx}{\sqrt{1-x^2}} = \frac{1\cdot 3}{2\cdot 4}\cdot\frac{\pi}{2} \qquad\qquad \int_0^1 \frac{x^5\,dx}{\sqrt{1-x^2}} = \frac{2\cdot 4}{3\cdot 5}$$

$$\int_0^1 \frac{x^6\,dx}{\sqrt{1-x^2}} = \frac{1\cdot 3\cdot 5}{2\cdot 4\cdot 6}\cdot\frac{\pi}{2} \qquad\qquad \int_0^1 \frac{x^7\,dx}{\sqrt{1-x^2}} = \frac{2\cdot 4\cdot 6}{3\cdot 5\cdot 7}$$

&c. &c.

or, referring to the general expressions at pages 45 and 46, we have, when m is an even number $= 2n$,

$$\int_0^1 \frac{x^{2n}\,dx}{\sqrt{1-x^2}} \quad \frac{1\cdot 3\cdot 5\cdot 7\ldots\ldots(2n-1)}{2\cdot 4\cdot 6\cdot 8\ldots\ldots\ldots 2n}\cdot\frac{\pi}{2},$$

and, when m is an odd number $= 2n + 1$

$$\int_0^1 \frac{x^{2n+1}\,dx}{\sqrt{1-x^2}} = \frac{2\cdot 4\cdot 6\cdot 8\cdots\cdots 2n}{3\cdot 5\cdot 7\cdot 6\ldots\ldots(2n+1)}.$$

If n is a number infinitely great, then will

$$\int_0^1 \frac{x^{2n}\,dx}{\sqrt{1-x^2}} = \int_0^1 \frac{x^{2n+1}\,dx}{\sqrt{1-x^2}};$$

hence, by division,

$$1 = \frac{\dfrac{2\cdot 4\cdot 6\cdot 8\cdot 10\ldots\ldots}{3\cdot 5\cdot 7\cdot 9\cdot 11\ldots\ldots}}{\dfrac{1\cdot 3\cdot 5\cdot 7\cdot 9\ldots\ldots}{2\cdot 4\cdot 6\cdot 8\cdot 10\ldots\ldots}}\cdot\frac{\pi}{2}$$

and, consequently,

$$\frac{\pi}{2} = \frac{\dfrac{2\cdot 4\cdot 6\cdot 8\cdot 10\ldots}{3\cdot 5\cdot 7\cdot 9\cdot 11\ldots}}{\dfrac{1\cdot 3\cdot 5\cdot 7\cdot 9\ldots}{2\cdot 4\cdot 6\cdot 8\cdot 10\ldots}} = \frac{2\cdot 2\cdot 4\cdot 4\cdot 6\cdot 6\cdot 8\cdot 8\ldots\ldots}{1\cdot 3\cdot 3\cdot 5\cdot 5\cdot 7\cdot 7\cdot 9\cdot 9\ldots}$$

a remarkable expression for the rectification of the circle first given by *Wallis.*

Modern English authors frequently put the above expression of Wallis in a very improper form, by writing it thus:

$$\frac{2^2 \cdot 4^2 \cdot 6^2 \dots \text{ad inf.}}{1 \cdot 3^2 \cdot 5^2 \dots \text{ad inf.}},$$

or thus:

$$\frac{2^2 \cdot 4^2 \cdot 6^2 \dots \text{ad inf.}}{3^2 \cdot 5^2 \cdot 7^2 \dots \text{ad inf.}},$$

neither of which expressions can represent the quadrant of a circle, for the first is infinite, and the second is 0.

3. To determine

$$\int_0^1 \frac{dx}{\sqrt{x\,(1 - x^2)}}.$$

$$\int_0^1 \frac{dx}{\sqrt{x(1 - x^2)}} = \left\{ 1 - \left(\frac{1}{2}\right)^2 + \left(\frac{1 \cdot 3}{2 \cdot 4}\right)^2 - \left(\frac{1 \cdot 3 \cdot 5}{2 \cdot 4 \cdot 6}\right)^2 + \&c. \right\} \pi.$$

4. To determine

$$\int_0^1 \frac{dx}{\sqrt{1 - x^4}}.$$

$$\int_0^1 \frac{dx}{\sqrt{1 - x^4}} = \left\{ 1 - \left(\frac{1}{2}\right)^2 + \left(\frac{1 \cdot 3}{2 \cdot 4}\right)^2 - \left(\frac{1 \cdot 3 \cdot 5}{2 \cdot 4 \cdot 6}\right)^2 + \&c. \right\} \frac{\pi}{2}.$$

In the foregoing examples the integral between the limits is obtained from the general integral previously found. But series have been determined for approximating to the value of the integral between limits without first finding the general integral. The investigation and application of these series, although an inquiry of considerable importance, cannot be with propriety fully entered into in an elementary treatise like the present; we must therefore content ourselves with referring the inquiring student to more extensive works on the Calculus, as the large treatise of *Lacroix*, the second volume of *Jephson's Fluxional Calculus*, and *M. Levy's* able article on the *Integral Calculus*, in the *Encyclopædia Metropolitana.*

(58.) We shall occupy the remaining part of the present chapter with a few examples of the application of the Integral Calculus to the

Summation of Series.

1. Required the sum of the series

$$s = x + 2x^2 + 3x^3 + 4x^4 + \ldots + nx^n.$$

Multiplying by $\dfrac{dx}{x}$ we have

$$s\,\frac{dx}{x} = dx + 2x\,dx + 3x^2\,dx + \ldots nx^{n-1}\,dx.$$

Integrating this

$$\int s\,\frac{dx}{x} = x + x^2 + x^3 + \ldots + x^n = \frac{x - x^{n+1}}{1 - x} + C.$$

The differential of this equation is

$$s\,\frac{dx}{x} = \frac{dx - (n+1)\,x^n\,dx + nx^{n+1}\,dx}{(1 - x)^2},$$

from which there results

$$s = \frac{x - (n+1)\,x^{n+1} + nx^{n+2}}{(1 - x)^2}.$$

2. Required the sum of the infinite series

$$s = \frac{x^{n+1}}{n+1} + \frac{x^{n+3}}{2 \cdot 3\,(n+3)} + \frac{x^{n+5}}{2 \cdot 3 \cdot 4 \cdot 5\,(n+5)} +$$

$$\frac{x^{n+7}}{2 \cdot 3 \cdot 4 \cdot 5 \cdot 6 \cdot 7\,(n+7)} + \&c.$$

Since (*Diff. Calc.* p. 29,)

$$e^x = 1 + x + \frac{x^2}{2} + \frac{x^3}{2 \cdot 3} + \frac{x^4}{2 \cdot 3 \cdot 4} + \frac{x^5}{2 \cdot 3 \cdot 4 \cdot 5} + \&c.$$

$$e^{-x} = 1 - x + \frac{x^2}{2} - \frac{x^3}{2 \cdot 3} + \frac{x^4}{2 \cdot 3 \cdot 4} - \frac{x^5}{2 \cdot 3 \cdot 4 \cdot 5} + \&c.$$

we have, by taking half the difference,

$$\frac{1}{2}e^x - \frac{1}{2}e^{-x} = x + \frac{x^3}{2 \cdot 3} + \frac{x^5}{2 \cdot 3 \cdot 4 \cdot 5} + \frac{x^7}{2 \cdot 3 \cdot 4 \cdot 5 \cdot 6 \cdot 7} + \&c.$$

hence, multiplying by $x^{n-1}dx$, and integrating, we have the following expression for the sum of the proposed series, viz.

$$\frac{1}{2}\int e^x \cdot x^{n-1}\, dx - \frac{1}{2}\int e^{-x} x^{n-1}\, dx =$$

$$\frac{x^{n+1}}{n+1} + \frac{x^{n+3}}{2\cdot 3\,(n+3)} + \frac{x^{n+5}}{2\cdot 3\cdot 4\cdot 5\,(n+5)} +$$

$$\frac{x^{n+7}}{2\cdot 3\cdot 4\cdot 5\cdot 6\cdot 7\,(n+7)} + \&c. = s.$$

But (30) and (31)

$$\frac{1}{2}\int e^x x^{n-1}\, dx = \frac{1}{2}e^x\left\{x^{n-1} - (n-1)\,x^{n-2} + \right.$$

$$\left.(n-1)(n-2)\,x^{n-3} - \&c.\right\}$$

$$\frac{1}{2}\int e^{-x} x^{n-1}\, dx = -\frac{1}{2}e^{-x}\left\{x^{n-1} + (n-1)\,x^{n-2} + \right.$$

$$\left.(n-1)(n-2)\,x^{n-3} + \&c.\right\},$$

consequently

$$s = \frac{1}{2}e^x\left\{x^{n-1} - (n-1)\,x^{n-2} + (n-1)(n-2)\,x^{n-3} - \&c.\right\} +$$

$$\frac{1}{2}e^{-x}\left\{x^{n-1} + (n-1)\,x^{n-2} + (n-1)(n-2)x^{n-3} + \&c.\right\}$$

Suppose $n=2$, and $x=1$, then this equation gives $s = e-1$, therefore

$$\frac{1}{e} = \frac{1}{2\cdot 71828\ldots} = \frac{1}{3} + \frac{1}{2\cdot 3\cdot 5} + \frac{1}{2\cdot 3\cdot 4\cdot 5\cdot 7} +$$

$$\frac{1}{2\cdot 3\cdot 4\cdot 5\cdot 6\cdot 7\cdot 9} + \&c.$$

3. To determine the sum of the infinite series

$$s = \frac{1}{p+q} + \frac{1}{p+2q} + \frac{1}{p+3q} + \ldots \frac{1}{p+nq}.$$

Let each term of this series be multiplied by the corresponding term of the series

$$x^{\frac{p}{q}+1}, \quad x^{\frac{p}{q}+2}, \quad x^{\frac{p}{q}+3}, \&c.$$

and we shall have the new series

$$S = \frac{x^{\frac{p}{q}+1}}{p+q} + \frac{x^{\frac{p}{q}+2}}{p+2q} + \frac{x^{\frac{p}{q}+3}}{p+3q} + \&c.$$

which agrees with the proposed, when $x = 1$.

By differentiating this, we get

$$q\,dS = x^{\frac{p}{q}}\,dx + x^{\frac{p}{q}+1}\,dx + x^{\frac{p}{q}+2}\,dx + \&c.$$

$$= (1 + x + x^2 + x^3 + x^4 + \&c.)\,x^{\frac{p}{q}}\,dx$$

$$= \frac{x^{\frac{p}{q}}}{1-x}\,dx \therefore S = \frac{1}{q}\int \frac{x^{\frac{p}{q}}}{1-x}\,dx,$$

consequently

$$s = \frac{1}{q}\int_0^1 \frac{x^{\frac{p}{q}}}{1-x}\,dx,$$

which is a general expression for the sum of the series, $x = 0$ being the origin of the integral, or the value for which it vanishes.

Suppose $p = 0$, then the above integral is $\frac{1}{q}$ log. $0 = \infty$; hence

$$\frac{1}{q} + \frac{1}{2q} + \frac{1}{3q} + \frac{1}{4q} + \&c. = \infty,$$

whatever be the finite value of q.

4. To determine the sum of the infinite series

$$s = \frac{1}{p+q} - \frac{1}{p+2q} + \frac{1}{p+3q} - \&c.$$

By proceeding as in last example, we find

$$q\,dS = (1 - x + x^2 - x^3 + x^4 - \&c.)\,x^{\frac{p}{q}}\,dx = \frac{x^{\frac{p}{q}}}{1+x}\,dx$$

$$\therefore S = \frac{1}{q}\int \frac{x^{\frac{p}{q}}}{1+x}\,dx \therefore s = \frac{1}{q}\int_0^1 \frac{x^{\frac{p}{q}}}{1+x}\,dx.$$

If we suppose $p = 0$, this integral is $\frac{1}{q}$ log. 2, therefore

K 2

$$\frac{1}{q}\left(1-\frac{1}{2}+\frac{1}{3}-\frac{1}{4}+\&c.\right)=\frac{1}{q}\log.\,2,$$

which we already know from other principles. (See the Essay on Logarithms, p. 3.)

5. To determine the sum of the infinite series

$$s=\frac{1}{(p+q)\,m}\pm\frac{1}{(p+2q)\,m^2}+\frac{1}{(p+3q)\,m^3}\pm\&c.$$

Let each term in this series be multiplied by the corresponding term in the series

$$x^{\frac{p}{q}+1},\;x^{\frac{p}{q}+2},\;x^{\frac{p}{q}+3},\;\&c.$$

and we shall have the new series

$$S=\frac{x^{\frac{p}{q}+1}}{(p+q)\,m}\pm\frac{x^{\frac{p}{q}+2}}{(p+2q)\,m^2}+\frac{x^{\frac{p}{q}+3}}{(p+3q)\,m^3}\pm\&c.$$

which will agree with the proposed, when $x=1$.

By differentiating this we have

$$qdS=\frac{x^{\frac{p}{q}}\,dx}{m}\pm\frac{x^{\frac{p}{q}+1}}{m^2}+\frac{x^{\frac{p}{q}+2}}{m^3}\pm\&c.$$

$$=\left(\frac{1}{m}\pm\frac{x}{m^2}+\frac{x^2}{m^3}\pm\&c.\right)x^{\frac{p}{q}}\,dx$$

$$=\frac{x^{\frac{p}{q}}}{m\pm x}\,dx\;\therefore\;S=\frac{1}{q}\int\frac{x^{\frac{p}{q}}}{m\mp x}\,dx$$

$$\therefore\;s=\frac{1}{q}\int_1^0\frac{x^{\frac{p}{q}}}{m\mp x}\,dx$$

6. To determine the sum of the infinite series whose general term is

$$\frac{1}{(p+qn)\,(r+sn)\,(t+un)\,\&c.}$$

n being the index of the term, or the number of its place in the series.

By examples 3 and 4

$$X_1 = \frac{1}{q}\int \frac{x^{\frac{p}{q}}}{1 \mp x}\,dx = \frac{x^{\frac{p}{q}+1}}{p+q} \pm \frac{x^{\frac{p}{q}+2}}{p+2q} + \frac{x^{\frac{p}{q}+3}}{p+3q} - \pm \&c.$$

Multiplying this equation by

$$x^{\frac{r}{s}-\frac{p}{q}-1}\,dx,$$

and integrating, we have

$$X_2 = \frac{1}{s}\int X_1 x^{\frac{r}{s}-\frac{p}{q}-1}\,dx = \frac{x^{\frac{r}{s}+1}}{(p+q)(r+s)} \pm$$

$$\frac{x^{\frac{r}{s}+2}}{(p+2q)(r+2s)} + \&c.$$

Multiplying this by

$$x^{\frac{t}{u}-\frac{r}{s}-1}\,dx,$$

and integrating, we have

$$X_3 = \frac{1}{u}\int X_2\, x^{\frac{t}{u}-\frac{r}{s}-1}\,dx = \frac{x^{\frac{t}{u}+1}}{(p+q)(r+s)(t+u)} \pm$$

$$\frac{x^{\frac{t}{u}+2}}{(p+2q)(r+2s)(t+2u)} + \&c.$$

and, by continuing this process, we shall obviously at length have, after m integrations, an expression X_m for the sum of a series of the kind proposed, of which the denominator of each term has m factors; observing to take the final integral between the limits $x = 0$ and $x = 1$.

If there are but two factors in the denominator of each term, that is, if the series is

$$\frac{1}{(p+q)(r+s)} \pm \frac{1}{(p+2q)(r+2s)} + \frac{1}{(p+3q)(r+3s)} + \&c.$$

then the general expression for the sum is

$$X_2 = \frac{1}{s} \int X_1\, x^{\frac{r}{s} - \frac{p}{q} - 1}\, dx = S$$

$$= \frac{1}{qs} \int x^{\frac{r}{s} - \frac{p}{q} - 1}\, dx \int \frac{x^{\frac{p}{q}}}{1 \mp x}\, dx,$$

which, by integrating by parts, becomes

$$\frac{x^{\frac{r}{s} - \frac{p}{q}}}{qs \left(\frac{r}{s} - \frac{p}{q}\right)} \int \frac{x^{\frac{p}{q}}}{1 \mp x}\, dx - \frac{1}{qs \left(\frac{r}{s} - \frac{p}{q}\right)} \int \frac{x^{\frac{r}{s}}}{1 \mp x}\, dx.$$

Now for $x = 1$ the coefficient of the first of these integrals becomes the same as that of the second; hence, between the limits $x = 1$ and $x = 0$, we have

$$s = \frac{1}{qs \left(\frac{r}{s} - \frac{p}{q}\right)} \int_0^1 \frac{x^{\frac{p}{q}} - x^{\frac{r}{s}}}{1 \mp x}\, dx \quad \ldots \ldots \text{(A)}.$$

When this general formula is applied to a series whose terms are all positive, then, it may be observed, the sum will be expressed purely algebraically provided $\dfrac{r}{s} - \dfrac{p}{q}$ be a whole number. For, putting $\dfrac{r}{s} - \dfrac{p}{q} = m$, the formula becomes

$$s = \frac{1}{qsm} \int_0^1 \frac{(1 - x^m)\, x^{\frac{p}{q}}}{1 - x}\, dx,$$

which is obviously algebraical, when m is a whole number, since $1 - x^m$ is divisible by $1 - x$ (*Alg.* p. 162.) But, if the signs of the terms are alternately positive and negative, then, that the sum may be algebraical, m must be an even whole number, for in this case the formula is

$$s = \frac{1}{qsm} \int_0^1 \frac{(1 - x^m)\, x^{\frac{p}{q}}}{1 + x}\, dx,$$

and $1 - x^m$ is not divisible by $1 + x$, unless m is even.

We shall now apply the general formula (A) to one or two particular cases, where the summation cannot be effected in algebraical terms, because, when the series is summable algebraically, the new and easy method explained in the Algebra is, we think, preferable as well on the ground of its greater simplicity as of its greater generality.

Required the sum of the infinite series

$$s = \frac{1}{1 \cdot 2} + \frac{1}{3 \cdot 4} + \frac{1}{5 \cdot 6} + \&c.$$

which agrees with the proposed form.

Here $p = -1$, $q = 2$, $r = 0$, $s = 2$, and for these values the formula (A) becomes

$$s = \frac{1}{2} \int_0^1 \frac{x^{-\frac{1}{2}} - 1}{1 - x}\, dx,$$

or, putting y for $x^{\frac{1}{2}}$

$$s = \int_0^1 \frac{dy}{1 + y} = \log. 2.$$

Required the sum of the series

$$s = \frac{1}{1 \cdot 4} - \frac{1}{2 \cdot 5} + \frac{1}{3 \cdot 6} - \frac{1}{4 \cdot 7} + \&c.$$

Here $p = 0$, $q = 1$, $r = 3$, $s = 1$, so that in this case the formula is

$$s = \frac{1}{3} \int_0^1 \frac{dx}{1 + x} - \frac{1}{3} \int_0^1 \frac{x^3}{1 + x}\, dx,$$

the general integrals are

$$\log. (1 + x), \text{ and } \frac{x^3}{3} - \frac{x^2}{2} + \cdots x \log. (1 + x)$$

$$\therefore s = \frac{2}{3} \log. 2 - \frac{5}{18}.$$

Required the sum of the infinite series

$$s = \frac{1}{1 \cdot 3} - \frac{1}{3 \cdot 5} + \frac{1}{5 \cdot 7} - \&c.$$

Here $p = -1$, $q = 2$, $r = 1$, $s = 2$, therefore the formula is

$$s = \frac{1}{4} \int_0^1 \frac{x^{-\frac{1}{2}} \, dx}{1+x} - \frac{1}{4} \int_0^1 \frac{x^{\frac{1}{2}}}{1+x} \, dx,$$

or, putting y for $x^{\frac{1}{2}}$

$$s = \frac{1}{2} \int_0^1 \frac{dy}{1+y^2} - \frac{1}{2} \int_0^1 \frac{y^2}{1+y^2} \, dy,$$

$$= \frac{1}{2} \int_0^1 \frac{dy}{1+y^2} - \frac{y}{2} - \frac{1}{2} \int_0^1 \frac{dy}{1+y^2}$$

$$= \int_0^1 \frac{dy}{1+y^2} - \frac{1}{2} = \frac{\pi}{4} - \frac{1}{2}.$$

We might now proceed to deduce the general expression for the series, when there are three factors in the denominator of each term, and then when there are four factors, and so on; but all this would occupy much more space than can be devoted here to these matters. We must refer, therefore, for these particulars to Clarke's translation of *Lorgna's* Method of Series. Without, however, deducing formulas for the summation of the various classes of series, included in the very comprehensive form proposed in the present example, we may obviously apply at once to any particular series the process of Lorgna, above exhibited, and it is this indeed that is usually done. Let it be required, for instance, to sum the infinite series

$$s = \frac{1}{1 \cdot 2 \cdot 4} - \frac{1}{2 \cdot 3 \cdot 5} + \frac{1}{3 \cdot 4 \cdot 6} + \&c.$$

Here we know that

$$X_1 = \int \frac{dx}{1+x} = x - \frac{x^2}{2} + \frac{x^3}{3} - \frac{x^4}{4} + \&c.$$

Multiplying by dx, and integrating,

$$X_2 = \int dx \int \frac{dx}{1+x} = \frac{x^2}{1 \cdot 2} - \frac{x^3}{2 \cdot 3} + \frac{x^4}{3 \cdot 4} - \&c.$$

Multiplying by xdx, and integrating

$$X_3 = \int xdx \int dx \int \frac{dx}{1+x} = \frac{x^4}{1 \cdot 2 \cdot 4} - \frac{x^5}{2 \cdot 3 \cdot 5} + \frac{x^6}{3 \cdot 4 \cdot 6} - \&c.$$

Now, by the integration by parts,

$$\int dx \int \frac{dx}{1+x} = x \log. (1 + x) - \int \frac{x}{1+x} dx = $$

$$(x + 1) \log. (1 + x) - x$$

$$\therefore X_3 = \int x (x + 1) dx \cdot \log. (1 + x) - \int x^2 dx = $$

$$(\frac{x^3}{3} + \frac{x^2}{2}) \log. (1 + x) - \frac{x^3}{3} - \int (\frac{x^3}{3} + \frac{x^2}{2}) \frac{dx}{1+x}$$

$$= (\frac{x^3}{3} + \frac{x^2}{2} - \frac{1}{6}) \log. (1 + x) - \frac{4x^3}{9} - \frac{x^2}{12} + \frac{x}{6}.$$

The results of these integrations need no correction, for they all vanish when $x = 0$, as they ought, since then X_1, X_2, X_3, are each 0. For $x = 1$ the last expression becomes

$$s = \frac{2}{3} \log 2 - \frac{19}{36},$$

which is the sum of the proposed series.

We shall now pass to other methods.

7. Required the sum of the infinite series

$$\frac{3^2}{4^2 \cdot 6} + \frac{3^2 \cdot 5^2}{4^2 \cdot 6^2 \cdot 8} + \frac{3^2 \cdot 5^2 \cdot 7^2}{4^2 \cdot 6^2 \cdot 8^2 \cdot 10} + \&c.$$

By example 2, page 96,

$$\int_0^1 \frac{x^{2n} dx}{\sqrt{1 - x^2}} = \frac{1 \cdot 3 \cdot 5 \dots (2n - 1)}{2 \cdot 4 \cdot 6 \dots \dots \dots 2n} \cdot \frac{\pi}{2}$$

Hence, if we assume

$$\frac{\pi}{2} s = \frac{2 \cdot 3}{4 \cdot 6} \int \frac{x^4 dx}{\sqrt{1 - x^2}} + \frac{2 \cdot 3 \cdot 5}{4 \cdot 6 \cdot 8} \int \frac{x^6 dx}{\sqrt{1 - x^2}} + $$

$$\frac{2 \cdot 3 \cdot 5 \cdot 7}{4 \cdot 6 \cdot 8 \cdot 10} \int \frac{x^8 dx}{\sqrt{1 - x^2}} + \&c.$$

and take the integrals between the limits $x = 0$ and $x = 1$, we shall have the sum of the proposed series equal to s.

By differentiating we get

$$\frac{\pi}{2} \cdot \frac{ds}{dx} \sqrt{1-x^2} = \frac{2 \cdot 3x^4}{4 \cdot 6} + \frac{2 \cdot 3 \cdot 5x^6}{4 \cdot 6 \cdot 8} + \frac{2 \cdot 3 \cdot 5 \cdot 7x^8}{4 \cdot 6 \cdot 8 \cdot 10} + \&c.$$

or, multiplying by x^2, and dividing by 4, we have

$$\frac{\pi}{2} \cdot \frac{ds}{dx} \cdot \frac{x^2 \sqrt{1-x^2}}{4} = \frac{3x^6}{2 \cdot 4 \cdot 6} + \frac{3 \cdot 5x^8}{2 \cdot 4 \cdot 6 \cdot 8} + \frac{3 \cdot 5 \cdot 7x^{10}}{2 \cdot 4 \cdot 6 \cdot 8 \cdot 10} + \&c.$$

but we know that

$$\sqrt{1-x^2} = 1 - \frac{x^2}{2} - \frac{x^4}{2 \cdot 4} - \frac{3x^6}{2 \cdot 4 \cdot 6} - \frac{3 \cdot 5x^8}{2 \cdot 4 \cdot 6 \cdot 8} -$$

$$\frac{3 \cdot 5 \cdot 7x^{10}}{2 \cdot 4 \cdot 6 \cdot 8 \cdot 10} - \&c.$$

hence, by addition,

$$\frac{\pi}{2} \cdot \frac{ds}{dx} \cdot \frac{x^2 \sqrt{1-x^2}}{4} + \sqrt{1-x^2} = 1 - \frac{x^2}{2} - \frac{x^4}{2 \cdot 4},$$

and, consequently,

$$\frac{\pi}{2} ds = \frac{4dx}{x^2 \sqrt{1-x^2}} - \frac{2dx}{\sqrt{1-x^2}} - \frac{x^2 dx}{2 \sqrt{1-x^2}} - \frac{4dx}{x^2};$$

therefore, integrating the differentials on the right, we find for their sum the expression

$$\frac{4 - 4 \sqrt{1-x^2}}{x} - 2\frac{1}{4}\sin.^{-1}x + \frac{1}{4} x \sqrt{1-x^2},$$

which between the proposed limits $x = 0$, $x = 1$, gives

$$\frac{\pi}{2} s = 4 - \frac{1}{4} \cdot \frac{\pi}{2} \therefore s = \frac{8}{\pi} - 2\frac{1}{4} = \cdot 296479\ldots$$

This question is taken from *Leybourn's Repository*, No. 20, and the following elegant solution to it is given by *Mr. Mason*, in No. 22 of the same valuable work.

By development

$$\frac{1}{\sqrt{1-x^2}} = 1 + \frac{x^2}{2} + \frac{3x^4}{2 \cdot 4} + \frac{3 \cdot 5x^6}{2 \cdot 4 \cdot 6} + \&c.$$

Multiply by dx, and integrate, and we have

$$\sin.^{-1} x = x + \frac{x^3}{2 \cdot 3} + \frac{3x^5}{2 \cdot 4 \cdot 5} + \frac{3 \cdot 5x^7}{2 \cdot 4 \cdot 6 \cdot 7} + \&c.$$

Multiply this by $\dfrac{xdx}{\sqrt{1-x^2}}$, and take the integrals on both sides between the limits $x=0$, $x=1$, and there results

$$1 = \frac{1}{2} \cdot \frac{\pi}{2} + \frac{1}{2^2 \cdot 4} \cdot \frac{\pi}{2} + \frac{3^2}{2^2 \cdot 4^2 \cdot 6} \cdot \frac{\pi}{2} +$$

$$\frac{3^2 \cdot 5^2}{2^2 \cdot 4^2 \cdot 6^2 \cdot 8} \cdot \frac{\pi}{2} + \&c.$$

Hence

$$\frac{8}{\pi} - 2\tfrac{1}{4} = \frac{3^2}{4^2 \cdot 6} + \frac{3^2 \cdot 5^2}{4^2 \cdot 6^2 \cdot 8} + \frac{3^2 \cdot 5^2 \cdot 7^2}{4^2 \cdot 6^2 \cdot 8^2 \cdot 10} + \&c.$$

8. Required the sum of the infinite series

$$\frac{1}{1^2 \cdot 3^2 \cdot 5^2 \ldots n^2} + \frac{1}{3^2 \cdot 5^3 \cdot 7^2 \ldots (n+2)^2} +$$

$$\frac{1}{5^2 \cdot 7^2 \cdot 9^2 \ldots (n+4)^4} + \&c.$$

n being any odd number whatever.

By (34)

$$\int_{\frac{\pi}{2}}^{0} \sin.^{n} x \, dx = \frac{2 \cdot 4 \cdot 6 \ldots n - 1}{3 \cdot 5 \cdot 7 \ldots n}.$$

Hence, if we assume

$$s = \frac{\int \sin.^{n} x \, dx}{2 \cdot 3 \cdot 4 \ldots n} + \frac{\int \sin.^{n+2} x \, dx}{2 \cdot 3 \cdot 4 \ldots (n+2)} + \frac{3^2 \int \sin.^{n+4} x \, dx}{2 \cdot 3 \cdot 4 \ldots (n+4)} +$$

$$\frac{3^2 \cdot 5^2 \int \sin.^{n+6} x \, dx}{2 \cdot 3 \cdot 4 \ldots (n+6)} + \&c.$$

and take the integrals between the limits $x=0$, $x=\dfrac{\pi}{2}$, s will be the sum of the proposed series.

By differentiating, we get

$$\frac{ds}{dx} = \frac{\sin.^n x}{2 \cdot 3 \cdot 4 \ldots n} + \frac{\sin.^{n+2} x}{2 \cdot 3 \cdot 4 \ldots (n+2)} + \frac{3^2 \sin.^{n+4} x}{2 \cdot 3 \cdot 4 \ldots (n+4)} +$$

$$\frac{3^2 \cdot 5^2 \cdot \sin.^{n+6} x}{2 \cdot 3 \cdot 4 \ldots (n+6)} + \&c.$$

and if we differentiate this result $n - 1$ times successively, we shall have

$$d^{n-1}\left(\frac{ds}{dx}\right) = \left\{ \sin. x + \frac{\sin.^3 x}{2 \cdot 3} + \frac{3^2 \sin.^5 x}{2 \cdot 3 \cdot 4 \cdot 5} + \frac{3^2 \cdot 5^2 \sin.^7 x}{2 \cdot 3 \cdot 4 \cdot 5 \cdot 6 \cdot 7} + \right.$$

$$\left. \&c. \right\} (d \sin. x)^{n-1}.$$

Now the series within the brackets is known to be equal to x, (*Diff. Calc.* p. 37); hence, by integrating,

$$\frac{ds}{dx} = \int^{n-1} x \, (d \sin. x)^{n-1}$$

$$\therefore s = \int_{\frac{\pi}{2}}^{0} \int^{n-1} x \, (d \sin. x)^{n-1} \, dx,$$

which is the general expression for the sum of the proposed series.

Suppose $n = 1$, then

$$s = \int_{\frac{\pi}{2}}^{0} x \, dx = \frac{1}{2} \left(\frac{\pi}{2}\right)^2 = 1 + \frac{1}{3^2} + \frac{1}{5^2} + \frac{1}{7^2} + \&c.$$

Again, let $n = 3$, then

$$s = \int_{\frac{\pi}{2}}^{0} \int^2 x \, (d \sin. x)^2 \, dx = \int_{\frac{\pi}{2}}^{0} \int (x \sin. x +$$

$$\cos. x - 1) \, d \sin. x \, dx =$$

$$\int_{\frac{\pi}{2}}^{0} \left(\frac{1}{2} x \sin.^2 x + \frac{3}{4} \sin. x \cos. x + \frac{1}{4} x - \sin. x\right) dx =$$

$$\frac{1}{4} \left(\frac{\pi}{2}\right)^2 - \frac{1}{2}$$

$$\therefore \ \frac{1}{4}\left(\frac{\pi}{2}\right)^2 - \frac{1}{2} = \frac{1}{1^2 \cdot 3^2} + \frac{1}{3^2 \cdot 5^2} + \frac{1}{5^2 \cdot 7^2} + \&c.$$

and so on.*

9. Required the sum of the infinite series

$$s = \frac{1}{3} - \frac{2}{4} + \frac{3}{5} - \frac{4}{6} + \&c.$$

$$s = \frac{3}{2} - 2\log. 2.$$

10. Required the sum of the infinite series

$$s = \frac{1}{1 \cdot 4} - \frac{1}{3 \cdot 6} + \frac{1}{5 \cdot 8} - \&c.$$

$$s = \frac{\pi}{12} + \frac{1}{6}\log. 2 - \frac{1}{6}.$$

11. Required the sum of the infinite series

$$s = \frac{1}{2 \cdot 2} - \frac{1}{3 \cdot 4} + \frac{1}{4 \cdot 8} - \&c.$$

$$s = 1 + 2\log. 2 - 2\log. 3.\dagger$$

12. Required the sum of the infinite series

$$\frac{1}{2^2 \cdot 4^2} + \frac{1^2}{2^2 \cdot 4^2 \cdot 6^2} + \frac{1^2 \cdot 3^2}{2^2 \cdot 4^2 \cdot 6^2 \cdot 8^2} + \frac{1^2 \cdot 3^2 \cdot 5^2}{2^2 \cdot 4^2 \cdot 6^2 \cdot 8^2 \cdot 10^2} + \&c.$$

$$s = \frac{32}{27\pi} - \frac{13}{36}.$$

---

* The mode of solution employed in this and in the former example, and which consists mainly in assimilating the proposed series to a series of integrals, taken between limits and multiplied by constant factors, is of extensive application, and will be found to succeed in many classes of series too complicated to be readily summed by the usual methods. It is proper to mention here that *Mr. Woolhouse*, of *North Shields*, was the first, as far as I know, who applied this very general and elegant method to series, and that the above example was proposed by him in the *Ladies' Diary* for 1830, shortly after the publication of which I forwarded the above solution to the Editor.

† For the summation of a great variety of other series, see *Clarke's*

SECTION II.

ON RECTIFICATION, QUADRATURE, AND CUBATURE.

CHAPTER I.

ON THE RECTIFICATION OF PLANE CURVES.

(59.) It has been shewn in the differential calculus page 124, that if s represent any arc of a plane curve, then

$$\frac{ds}{dx} = \sqrt{1 + \frac{dy^2}{dx^2}}$$

or

$$ds = \sqrt{1 + \frac{dy^2}{dx^2}} \cdot dx,$$

x and y representing the coordinates of one of its extremities. Hence to determine the general relation between s and x, we must integrate this expression, so that

$$s = \int \sqrt{1 + \frac{dy^2}{dx^2}} \cdot dx$$

or, if we interchange the axes to which the curve is referred,

$$s = \int \sqrt{1 + \frac{dx^2}{dy^2}} \cdot dy.$$

Either of these expressions may be considered as a general formula for the length s of any arc of any plane curve, referred to rectangular coor-

---

translation of Lorgna; Wright's Solutions to the Cambridge Problems, vol. 1; and *Young's Treatise on Algebra.*

dinates, $\dfrac{dy}{dx}$ being a function of x, and $\dfrac{dx}{dy}$ a function of y, equally dependent on the equation of the curve.

If we take any proposed curve, and substitute in either of these formulas, instead of the general symbols $\dfrac{dy}{dx}, \dfrac{dx}{dy}$, the particular function of x or of y, given by the equation of this curve, the expressions for s will then become less general, since they will be restricted to the arcs of the proposed curve; but they will still be indefinite, since nothing as yet fixes the arbitrary constant which they each involve. If, however, we fix upon any point in the proposed curve from which the arc is to be measured, then we at the same time fix the value of the constant, for at that point $s = 0$, so that the the expression for s belongs only to the arc commencing at the given point, and terminating at the point $(x, y.)$

EXAMPLES.

(60.) 1. To determine the length of an arc of a parabola measured from the vertex.

The equation of the curve is

$$y^2 = 4mx \therefore \frac{dx}{dy} = \frac{y}{2m};$$

hence

$$s = \int \sqrt{1 + \frac{dx^2}{dy^2}} \cdot dy = \frac{1}{2m} \cdot \int \sqrt{y^2 + 4m^2} \cdot dy$$

and, by (17),

$$\int \sqrt{y^2 + 4m^2} \cdot dy = \frac{y\sqrt{y^2 + 4m^2}}{2} + \frac{4m^2 \log. (y + \sqrt{y^2 + 4m^2})}{2} + C,$$

consequently

$$s = \frac{y\sqrt{y^2 + 4m^2}}{4m} + m \log. (y + \sqrt{y^2 + 4m^2}) + C.$$

Since the arc commences at the origin, therefore when $y = 0$, $s = 0$ that is

$$0 = m \log. 2m + C$$

$$\therefore C = - m \log. 2m$$

$$\therefore s = \frac{y \sqrt{y^2 + 4m^2}}{4m} + m \log. \left\{ \frac{y + \sqrt{y^2 + 4m^2}}{2m} \right\}$$

which expresses the length of any arc of the parabola measured from the vertex, in terms of the ordinate of its other extremity.

If instead of being supposed to commence at the origin the arc commenced at a point of which the ordinate is b, then the constant would be determined by the condition that $s = 0$ when $y = b$, which condition would give

$$C = - \left\{ \frac{b \sqrt{b^2 + 4m^2}}{4m} + m \log. (b + \sqrt{b^2 + 4m^2}) \right\}$$

so that the length of any arc measured from the point $y = b$ is

$$s = \frac{y \sqrt{y^2 + 4m^2} - b \sqrt{b^2 + 4m^3}}{4m} + m \log. \frac{y + \sqrt{y^2 + 4m^2}}{b + \sqrt{b^2 + 4m^2}},$$

which, when $b = 0$, becomes identical to the former expression, as it ought.

It appears from this example that the length of any arc of the common parabola may always be expressed in finite terms, although the arc is not in strictness rectifiable, since the expression for its length involves a transcendental quantity, which cannot be expressed numerically in finite terms. There are, however, an infinite number of parabolas of the higher orders which are completely rectifiable; we shall determine the general equation of these in the next example.

2. To determine the class of parabolas which are rectifiable.

The general equation of parabolas of all orders is

$$y^m = ax^n \quad \therefore \quad \frac{dy}{dx} = \frac{n}{m} x^{\frac{1}{m}} x^{\frac{n}{m} - 1}$$

$$\therefore s = \int \sqrt{1 + \frac{dy^2}{dx^2}} \cdot dx = \int \left\{ 1 + \frac{n^2}{m^2} a^{\frac{2}{m}} x^{2 \left(\frac{n}{m} - 1 \right)} \right\}^{\frac{1}{2}}$$

By referring to the criterion (19), we find that for this integral to be wholly algebraical and finite, we must have the condition

$$1 \div 2 \left(\frac{n}{m} - 1\right) = \text{a positive whole number,}$$

therefore, calling this whole number w, we have

$$\frac{n}{m} = \frac{1 + 2w}{2w},$$

so that the curve will always be rectifiable when one of the exponents is an even number $(2w)$ and the other exceeds it by unity, or when they are equimultiples of such numbers.

3. To rectify the circle.

The equation, accordingly as we assume the origin at the centre or at the circumference, will be

$$y^2 = r^2 - x^2 \text{ or } y^2 = 2rx - x^2,$$

and the expression for s will therefore be either

$$s = r \int \frac{dx}{\sqrt{r^2 - x^2}} \text{ or } s = r \int \frac{dx}{\sqrt{2rx - x^2}},$$

both of which involve circular arcs. The circle is not therefore a rectifiable curve. Either of these integrals may, however, be developed into a series, and thus an approximation to the circumference obtained, but a very convergent series for this purpose has already been investigated in the Differential Calculus, page 37.

(61.) 4. To rectify the ellipse

The equation of the curve is (*Anal. Geom.* p. 101,)

$$y^2 = (\epsilon^2 - 1)(x^2 - a^2)$$

$$\therefore \frac{dy}{dx} = \frac{x\sqrt{\epsilon^2 - 1}}{\sqrt{x^2 - a^2}} \therefore s = \int \sqrt{1 + \frac{x^2(\epsilon^2 - 1)}{x^2 - a^2}} \, dx$$

$$= \int \frac{\sqrt{a^2 - \epsilon^2 x^2}}{\sqrt{a^2 - x^2}} \, dx = a \int \frac{\sqrt{1 - \epsilon^2 x'^2}}{\sqrt{1 - x'^2}} \, dx'.$$

x' being put for $\dfrac{x}{a}$.

This integration cannot be effected in finite terms, but the integration by series has already been given at length at page 86. Suppose a quadrant of the ellipse is required, then the arc to be rectified commences where $x = 0$, and terminates where $x = a$; and between these limits the series referred to becomes

$$\frac{\pi}{2}\left\{1 - \frac{1 \cdot 1}{2 \cdot 2}\,\varepsilon^2 - \frac{1 \cdot 1 \cdot 1 \cdot 3}{2 \cdot 2 \cdot 4 \cdot 4}\,\varepsilon^4 - \frac{1 \cdot 1 \cdot 1 \cdot 3 \cdot 3 \cdot 5}{2 \cdot 2 \cdot 4 \cdot 4 \cdot 6 \cdot 6}\,\varepsilon^6 - \&c.\right\},$$

which, therefore, expresses the length of the elliptic quadrant, $\frac{\pi}{2}$ being

the circular quadrant whose radius is a, the semi-major axis of the ellipse. Hence the whole periphery of the ellipse is found by multiplying the circumference of the circumscribing circle by the series within the brackets.

Suppose, for example, it were required to find the periphery of the ellipse whose semi axes are 12 and 9,

$$\varepsilon^2 = 1 - \frac{b^2}{a^2} = 1 - \frac{3^2}{4^2} = \cdot 4375$$

Then the second term, A, $= \dfrac{\varepsilon^2}{4} = 1 \cdot 10938$

third   B, $= \dfrac{3\varepsilon^2}{4 \cdot 4}$ A $= \cdot 00897$

fourth   C, $= \dfrac{3 \cdot 5\varepsilon^2}{6 \cdot 6}$ B $= \cdot 00164$

fifth   D, $= \dfrac{5 \cdot 7\varepsilon^2}{8 \cdot 8}$ C $= \cdot 00039$

sixth   E, $= \dfrac{7 \cdot 9\varepsilon^2}{10 \cdot 10}$ D $= \cdot 00011$

seventh   F, $= \dfrac{9 \cdot 11\varepsilon^2}{12 \cdot 12}$ E $= \cdot 00003$

eighth   G, $= \dfrac{11 \cdot 13\varepsilon^2}{14 \cdot 14}$ F $= \cdot 00001$

Sum $= \cdot 12053$

and this, taken from the first term of the series, 1, leaves $\cdot 87947$ which multiplied by $3 \cdot 1416 \times 24$, the circumference of the circumscribing circle, gives $66 . 31056$ for the periphery of the proposed ellipse.

The foregoing series for the rectification of the ellipse, which is that usually given, is not so convergent as might be wished when the ellipse differs but little from a circle, in which case ϵ differs but little from unity. We shall here, therefore, investigate another and more convergent series for this purpose.

If a circle be described on the major axis of an ellipse, and the ordinate of any point be produced to meet the circumference, then the abscissa x of the same point will be the cosine of the angle subtended by this line at the centre; hence, calling this angle ϕ and taking the tabular cosine, we have $x = a \cos. \phi$; hence

$$\frac{\sqrt{a^2 - \epsilon^2 x^2}}{\sqrt{a^2 - x^2}}\, dx = \frac{\sqrt{1 - \epsilon^2 \cos.^2 \phi}}{\sqrt{1 - \cos.^2 \phi}}\, a\, d \cos. \phi,$$

or since

$$d \cos. \phi = - \sin. \phi\, d\phi = - \sqrt{1 - \cos.^2 \phi}\, . \, d\phi$$

$$\frac{\sqrt{a^2 - \epsilon^2 x^2}}{\sqrt{a^2 - x^2}}\, dx = - a \sqrt{1 - \epsilon^2 \cos.^2 \phi}\, . \, d\phi,$$

or putting for $\cos.^2 \phi$ its equal $\frac{1}{2} + \frac{1}{2} \cos. 2\phi$ (*Gregory's Trig.* p. 46,) the expression for the differential of the elliptic arc becomes

$$- a \sqrt{1 - \frac{\epsilon^2}{2} - \frac{\epsilon^2}{2} \cos. 2\, \phi}\, . \, d\phi,$$

of which the integral will be s, the length of the arc. This expression will take a convenient form for development if we determine a' and b', so that it may be identical to

$$- a \sqrt{a'^2 + b'^2 - 2a'b' \cos. 2\phi}\, . \, d\phi.$$

The equations for determining a' and b' are

$$a'^2 + b'^2 = 1 - \frac{\epsilon^2}{2} \text{ and } 2a'b' = \frac{\epsilon^2}{2},$$

whence

$$a' + b' = 1,\ a' - b' = \sqrt{1 - \epsilon^2} \therefore a' =$$

$$\frac{1 + \sqrt{1 - \epsilon^2}}{2},\ b' = \frac{1 - \sqrt{1 - \epsilon^2}}{2}\, .$$

We have then to develop the expression above in a series of terms convenient for integration. Since *(Diff. Calc.* p. 30,)

$$2 \cos. \phi = e^{\phi\sqrt{-1}} + e^{-\phi\sqrt{-1}},$$

it follows that

$$a'^2 + b'^2 - 2a'b' \cos. \phi = (a' - b'e^{\phi\sqrt{-1}})(a' - b'e^{-\phi\sqrt{-1}};$$

therefore, developing the nth power of each factor by the binomial theorem, we have for

$$(a'^2 + b'^2 - 2a'b' \cos. \phi)^n$$

the product of the two series

$$a'^n \left\{ 1 - n\,\frac{b'}{a'}\, e^{\phi\sqrt{-1}} + \frac{n\,(n-1)}{1\cdot 2}\, e^{2\phi\sqrt{-1}} - \right.$$

$$\left. \frac{n\,(n-1)\,(n-2)}{1\cdot 2\cdot 3}\, e^{3\phi\sqrt{-1}} + \&c. \right\}$$

and

$$a'^n \left\{ 1 - n\,\frac{b'}{a'}\, e^{-\phi\sqrt{-1}} + \frac{n\,(n-1)}{1\cdot 2}\, e^{-2\phi\sqrt{-1}} - \right.$$

$$\left. \frac{n\,(n-1)\,(n-2)}{1\cdot 2\cdot 3}\, e^{-3\phi\sqrt{-1}} + \&c. \right\},$$

which product we find by putting $2 \cos. m\phi$ for its equal

$$e^{m\phi\sqrt{-1}} + e^{-m\phi\sqrt{-1}}$$

to be

$$a'^{2n} \left\{ 1 + n^2\,\frac{b'^2}{a'^2} + \frac{n^2\,(n-1)^2}{2^2}\,\frac{b'^4}{a'^4} + \frac{n^2\,(n-1)^2\,(n-2)^2}{2^2\cdot 3^2}\cdot\frac{b'^6}{a'^6} + \right.$$

$$\left. \&c. \right\} -$$

$$2a'^{2n} \left\{ n\,\frac{b'}{a'} + n\cdot\frac{n\,(n-1)}{2}\,\frac{b'^3}{a'^3} + \frac{n\,(n-1)}{2}\cdot\frac{n\,(n-1)\,(n-2)}{2\cdot 3}\,\frac{b'^5}{a'^5} + \right.$$

$$\left. \&c. \right\} \cos. \phi +$$

$$P \cos. 2\,\phi + Q \cos. 3\,\phi + R \cos. 4\,\phi + \&c.;$$

hence, multiplying by $- d\phi$ and taking the integrals of the several terms between the limits $\phi = \pi$ and $\phi = 0$, all vanish but the first term, the integral between the proposed limits being

$$\pi \, a'^{2u} \left\{ 1 + n^2 \frac{b'^2}{a'^2} + \frac{n^2 (n-1)^2}{2^2} \cdot \frac{b'^4}{a'^4} + \frac{n^2 (n-1)^2 (n-2)^2}{2^2 \cdot 3^2} \cdot \frac{b'^6}{a'^6} + \&c. \right\}$$

therefore when $n = \frac{1}{2}$ we have for the semi-periphery of the ellipse the series

$$\pi \, a' \left\{ 1 + \frac{1^2}{2^2} \cdot \frac{b'^2}{a'^2} + \frac{1^2 \cdot 1^2}{2^2 \cdot 4^2} \cdot \frac{b'^4}{a'^4} + \frac{1^2 \cdot 1^2 \cdot 3^2}{2^2 \cdot 4^2 \cdot 6^2} \cdot \frac{b'^6}{a'^6} + \&c. \right\}.$$

The coefficients in this series obviously converge faster than those in the series first given, and moreover

$$\frac{b'}{a'} = \frac{1 - \sqrt{1 - \varepsilon^2}}{1 + \sqrt{1 + \varepsilon^2}} = \frac{2 \sqrt{1 - \varepsilon^2} - 2}{\varepsilon^2} + 1$$

is necessarily always smaller than ε.

The foregoing investigation is taken with slight modification from *Mr. Ivory's* paper on the Rectification of the Ellipsis in the Edinburgh Phil. Trans., vol. iv.

A series for the rectification of the hyperbola may be obtained in a similar manner.

(62.) 5. A given circle rolls along a given straight line always remaining in the same plane; it is required to determine the length of the track described by any point P in its circumference?

The curve P, P′, P″, P‴, thus generated is called a *cycloid*, and in order to determine its length we must first find its differential equation.

Let P‴P, P‴Y be the axes of reference; then, taking any point P″ in the curve and the corresponding position of the generating circle, it is obvious that the straight line MP‴ must be equal to the arc MP″; hence, calling the tabular angle corresponding to this arc ω, we have

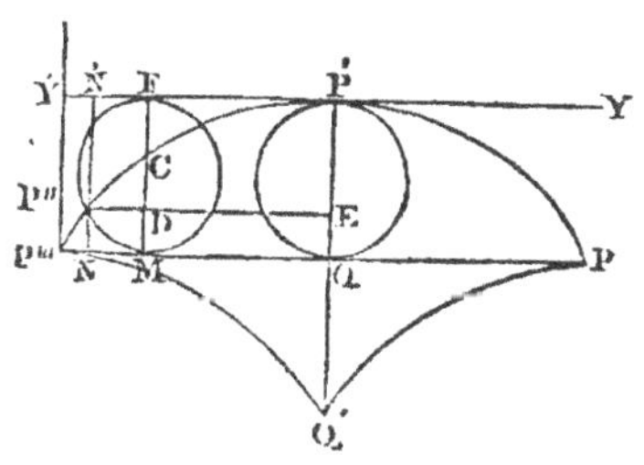

$$\mathrm{P'''\,N = P''\,M - NM,}$$

or

$$x = r\omega - r \sin. \omega \, \ldots \ldots \, (1),$$

also

$$\mathrm{P''\,N = CM - CD}$$

or

$$y = r - r \cos. \omega \ \ldots \ldots (2).$$

From (1),

$$dx = (r - r \cos. \omega) \, d\omega = y \, d\omega,$$

from (2),

$$dy = r \sin. \omega \, d\omega$$

$$\therefore \ \frac{dy}{dx} = \frac{r \sin. \omega}{y},$$

but since by the circle $r \sin. \omega$ or $P''D$ is equal to $\sqrt{2ry - y^2}$, we have, by substitution,

$$\frac{dy}{dx} = \frac{\sqrt{2ry - y^2}}{y} = \sqrt{\frac{2r - y}{y}}$$

for the differential equation of the cycloid. Hence

$$s = \int \sqrt{1 + \frac{dx^2}{dy^2}} \cdot dy = \int \sqrt{1 + \frac{y}{2r - y}} \cdot dy = \sqrt{2r} \int \frac{dy}{\sqrt{y}} =$$

Considering the curve to commence at P''' we have $s = 0$ when $y = 0$ $\therefore C = \cancel{0}.4^r$ From the commencement P''' to P' the middle of the curve, and at which point $y = 2r$, we have $s = 4r$, so that the whole length of the cycloid is equal to 4 times the diameter of its generating circle.

The equation of the cycloid in terms of x and y is very easily obtained as follows:

$$P'''N = P'''M - P''D = \sin.^{-1} P''D - P''D,$$

that is, since $P''D = \sqrt{2ry - y^2}$,

$$x = \sin.^{-1} \sqrt{2ry - y^2} - \sqrt{2ry - y^2},$$

or, which is the same thing,

$$x = \text{versin.}^{-1} y - \sqrt{2ry - y^2},$$

the radius of the arc being r.

Again

$$P''N = DM = r - CD = r - \cos. MP'',$$

that is, since $MP'' = DP'' + NP'''$,

$$y = r - \cos. (\sqrt{2ry - y^2} + x),$$

✳ Hence

$$S = 4r - 2\sqrt{2r(2r - y)}$$

which is another form of the equation, and either of these being differentiated will furnish the same differential equation as that above.

If the origin of the coordinates be at P′, the vertex of the curve, the
axes being P′Y, P′Q, the equation is found with equal ease, for since
the ordinate N′P″, which is always negative, is

$$N'P'' = FC + CD = -r - CD,$$

and CD = — cos. FP″; but FP″ = FP′ or MQ, since the whole
semicircle is equal to P′″Q or YP′, also FP′ = N′P′ — N′F; hence

$$N'P'' = -r + \cos. (N'P' - N'F)$$

or

$$y = -r + \cos. (x - \sqrt{-2ry - y^2}).$$

In this equation x is of course negative for that half of the cycloid
to which our reasoning here is applied, and it is positive for the other half.

It is a curious property of this curve that its evolute consists of two
inverted semi-cycloids, each equal to half the proposed cycloid. It
is worth while to prove this.

Representing as usual $\dfrac{dy}{dx}$ by p', we have, by differentiating the
expression (3) with respect to x,

$$\frac{dp'}{dy} \cdot \frac{dy}{dx} = p'' = -\frac{r}{y^2},$$

and if α, β, represent the coordinates of any point in the evolute
corresponding to $(x, y,)$ in the involute, we know (*Diff. Calc.* p. 139,)
that

$$\alpha = x - \frac{p'(p'^2 + 1)}{p''}, \ \beta = y + \frac{p'^2 + 1}{p''},$$

that is

$$\alpha = x + 2\sqrt{2ry - y^2}, \ \beta = -y,$$

from the second of these we get

$$x = \alpha - 2\sqrt{-2r\beta - \beta^2},$$

which values of x and y substituted in one of the foregoing equations
of the curve, the last for instance, give

$$\beta = -r + \cos. \left(-\sqrt{-2r\beta - \beta^2} + a \right),$$

and this equation agrees exactly with that of the proposed cycloid when the origin of the axes is removed from P''' to the vertex P'; hence, the evolute of the semi-cycloid $P'''P'$ is an equal cycloid, $P'''Q'$ having its vertex at P''', and consequently one extremity of its base at Q', $P'Q'$ being $= 2P'Q$; also, the curve being symmetrical with respect to the axis $P'Q'$, the evolute of PP' must be a semi-cycloid PQ' symmetrical with the former.

6. To determine the length of the arc of the parabola whose equation is $y^3 = nx^2$ between the limits $x = 0$, $x = a$:

$$s = \frac{8n}{27} \left\{ \frac{9}{4} \left(\frac{x}{n} \right)^{\frac{2}{3}} + 1 \right\}^{\frac{3}{2}} - \frac{8n}{27}.$$

7. Required the length of the curve whose equation is

$$y = (a^{\frac{2}{3}} - x^{\frac{2}{3}})^{\frac{3}{2}},$$

from $x = 0$ to $x = a$,

$$s = \frac{3}{2} a^{\frac{1}{3}} x^{\frac{2}{3}}.$$

8. Prove that in any plane curve the length of the tangent at any point (x, y) is

$$T = \pm y \frac{(ds)}{(dy)},$$

the independent variable being arbitrary. The upper sign obviously has place if both s and y encrease or decrease together, but the lower sign, if one, encreases while the other diminishes.

9. To rectify the *tractrix* of which the figure is given in the margin, and whose characteristic property is, that if from any point P in the curve a tangent be drawn, the part PE between the point and the axis AX is equal to the constant quantity $a = AY$,

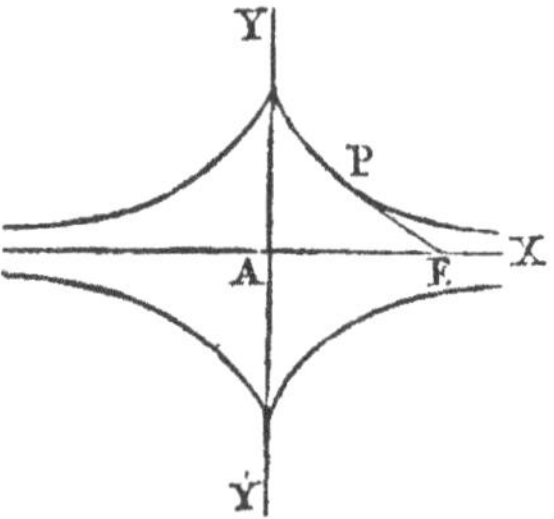

$$s = a \log. \frac{a}{y}.$$

10. To determine the length of the curve whose characteristic property is such that a line $x'p$ drawn from the foot of the ordinate, perpendicular to and terminated by the tangent PR, is equal to the constant quantity $a = \mathrm{AA}'$:

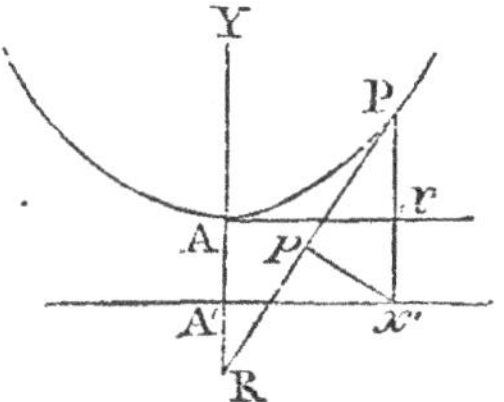

$$s = \sqrt{2ax + x^2}.$$

This is the *catenary* or curve formed by a heavy and perfectly flexible chain, suspended by its extremities

(63.) Let us now determine the formula for the rectification of a plane curve when it is referred to polar instead of rectangular coordinates.

We know by the formulas for changing the independent variable, (*Diff. Calc.* p. 98,) that $\dfrac{dy}{dx}$ is the same as $\dfrac{(dy)}{(dx)}$, the independent variable in this latter coefficient being any whatever, so that the formula at the head of this chapter when put in its most general form, as indeed we have given it at page 125 of the Differential Calculus, is

$$(ds) = \sqrt{(dx)^2 + (dy)^2};$$

therefore, when the angle ω between the radius vector and fixed axis is taken for the independent variable, the formula is

$$\frac{ds}{d\omega} = \sqrt{\frac{dx^2}{d\omega^2} + \frac{dy^2}{d\omega^2}};$$

but (*Diff. Calc.* p. 137,)

$$\frac{dy}{d\omega} = r \cos. \omega + \frac{dr}{d\omega} \sin. \omega, \frac{dx}{d\omega} = -r \sin. \omega + \frac{dr}{d\omega} \cos. \omega;$$

hence, by substituting these values in the foregoing expression, we have

$$s = \int \sqrt{r^2 + \frac{dr^2}{d\omega^2}} \cdot d\omega \ \ldots \ldots (1).$$

(64.) It is worthy of remark that the foregoing expression for $\dfrac{ds}{d\omega}$ is the same as that for the length of the polar normal PN; for the expression for the subnormal FN is $\dfrac{dr}{d\omega}$, (*Diff. Calc.* p. 118,) consequently

$$PN = \sqrt{FN^2 + FP^2} = \sqrt{r^2 + \frac{dr^2}{d\omega^2}}.$$

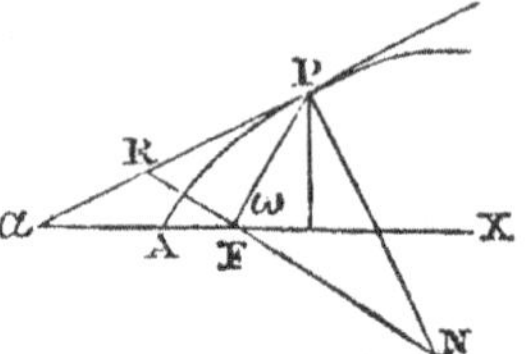

If from F we conceive a perpendicular to be demitted on the tangent PR, and call the part of the tangent intercepted between this perpendicular and the point of contact t, then

$$PN : FN :: FP : t,$$

that is

$$\frac{ds}{d\omega} : \frac{dr}{d\omega} :: r : t$$

$$\therefore \frac{ds}{d\omega} = \frac{r\,\dfrac{dr}{d\omega}}{t},$$

or, considering the independent variable as arbitrary, (*Diff. Calc.* p. 98,)

$$(ds) = \frac{r\,(dr)}{t} \;\ldots\ldots (2).$$

EXAMPLES.

(65.)   1. To determine the length of an arc of the logarithmic spiral.

The equation of this spiral is $r = a^{\omega}$,

$$\therefore \frac{dr}{d\omega} = \log. a \cdot a^{\omega} = \log. a \cdot r \therefore d\omega = \frac{dr}{r \log. a};$$

hence

$$s = \int \sqrt{r^2 + \frac{dr}{d\omega}} \cdot d\omega = \int \sqrt{1 + \frac{1}{\log.^2 a}} \cdot dr$$

$$= \sqrt{1 + \frac{1}{\log.^2 a}} \cdot r + C$$

$$= r \sec. \angle F + C; \text{ (see p. 118, } Diff. Calc.)$$

Ift he arc is to be measured from the pole, then the length between the pole and the point, whose distance from it is r, will be r sec. $\angle$ F; but if we require the length of the arc comprised between two points distant r' and r from the pole, the expression will be $(r - r')$ sec. F.

2. To determine the length of an arc of the spiral of Archimedes. The equation of this spiral is

$$r = a\omega \therefore \frac{dr}{d\omega} = a \therefore d\omega = \frac{dr}{a}$$

$$\therefore s = \frac{1}{a} \int \sqrt{r^2 + a^2} \cdot dr.$$

This integral is the same as that expressing the arc of a parabola found in art. 60; hence

$$s = \frac{r \sqrt{r^2 + a^2}}{2a} + \frac{a}{2} \log. \frac{r + \sqrt{r^2 + a^2}}{a} + C.$$

If the arc commence at the pole, then $s = 0$ when $r = 0 \therefore C = 0$.

3. To determine the length of the involute of the circle.

The involute may be described by the unwinding of a string from the circumference ABC; and since in every position CP it will be tangent to the circle and normal to the curve AP, (*Diff. Calc.* p. 141,) it follows that OR parallel to CP will be perpendicular to the tangent PR, and therefore RP = OC = a; hence, by equation (2) above,

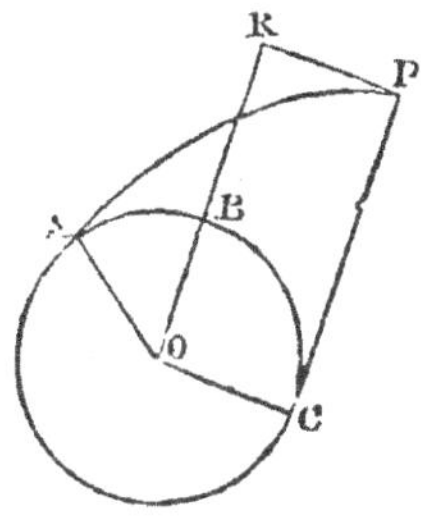

$$s = \frac{1}{a} \int r \, dr = \frac{r^2}{2a} + C.$$

Now when $r = a = $ OA, then $s = 0 \therefore C = -\frac{a}{2}$;

$$\therefore s = \frac{r^2 - a^2}{2a} = \frac{CP^2}{2OC},$$

the length of any arc AP.

4. To determine the length of an arc of the reciprocal spiral $r = \frac{1}{\omega}$, commencing at the pole

M 2

$$s = \sqrt{1 + r^2} + \log. \frac{r}{\sqrt{1 + r^2} + 1} - 1.$$

5. To determine the length of an arc of a curve whose polar equation is

$$r = 2a\,(1 + \cos. \omega);$$

the arc being comprised between two points at which $\omega = 0$ and $\omega = \pi$

$$s = 8a.$$

CHAPTER II.

ON THE QUADRATURE OF CURVES.

(66.) Let us now seek an expression for the area of any portion of a curve surface situated in a plane, and for this purpose let us first determine the differential expression for a plane surface.

In the plane curve surface ABCM′, any portion ABM is obviously a function of the coordinates AM MB, or simply of the abscissa AM, since both the area and the abscissa always vary together, and we are now to find the general expression for the differential of this function.

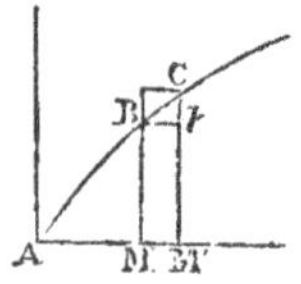

Take any increment, MM′ $= h$, of the abscissa, and draw the corresponding ordinate M′C, and then complete the parallelograms BM′, CM; bC will be the increment k of MB, and it is obvious that the increment CBMM′ of the surface is always between the two parallelograms however we diminish the increment MM′; if, therefore, the ratio of these parallelograms could ever be that of equality, the ratio of the corresponding curvilinear increment to either would also necessarily be that of equality.

The ratio of the parallelograms is always

$$\frac{yh}{(y+k)\,h} = \frac{y}{y+k},$$

and in the limit, that is, when $h = 0$ and consequently $k = 0$, it is simply

$$\frac{y}{y} = 1,$$

which being a ratio of equality it follows that in the limit

$$\frac{yh}{\text{CBMM}'} = 1,$$

that is

$$\frac{y\,dx}{d\,.\,\text{area}} = 1 \therefore d\,.\,\text{area} = y\,dx$$

$$\therefore \text{area} = \int y\,dx = u.$$

If the axes of coordinates were oblique, then, calling their included angle ϕ, the area of the parallelogram Mb would not be yh but $\sin.\phi yh$; hence, in that case, $d\,.\,\text{area} = \sin.\phi y dx$,

$$\therefore \text{area} = \sin.\phi \int y\,dx.$$

We shall now give a few applications of these formulas.

EXAMPLES.

(67.) 1. To determine the area of the parabola

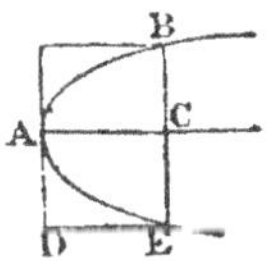

$$y^2 = ax \therefore 2y\,dy = a\,dx$$

$$\therefore u = \int y\,dx = \frac{2}{a} \int y^2\,dy = \frac{2}{3} \cdot \frac{y^3}{a} = \frac{2}{3} xy;$$

hence the area ABC is equal to two thirds of the parallelogram AB, or the whole area BAE equal to two thirds of the circumscribing parallelogram BD, so that this curve is accurately quadrable.

2. To determine the area of the circle

$$y = \sqrt{r^2 - x^2} \therefore \int y\,dx = \int \sqrt{r^2 - x^2}\,.\,dx,$$

developing the radical, we have

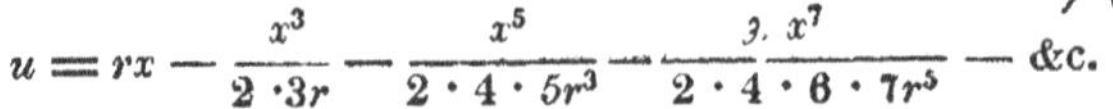

$$u = rx - \frac{x^3}{2 \cdot 3r} - \frac{x^5}{2 \cdot 4 \cdot 5r^3} - \frac{3 \cdot x^7}{2 \cdot 4 \cdot 6 \cdot 7r^5} - \&c.$$

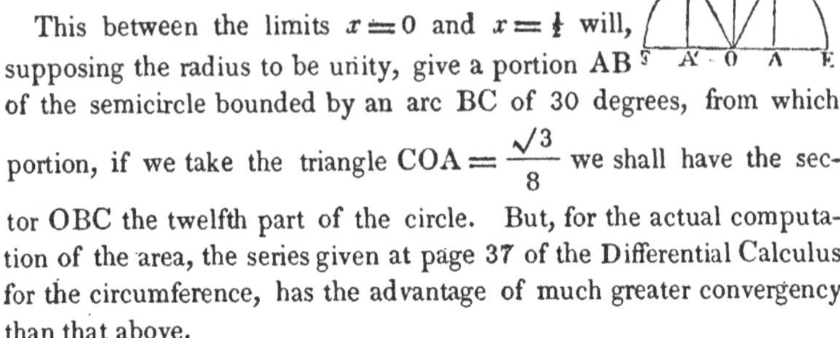

This between the limits $x = 0$ and $x = \frac{1}{2}$ will, supposing the radius to be unity, give a portion AB of the semicircle bounded by an arc BC of 30 degrees, from which portion, if we take the triangle $\mathrm{COA} = \dfrac{\sqrt{3}}{8}$ we shall have the sector OBC the twelfth part of the circle. But, for the actual computation of the area, the series given at page 37 of the Differential Calculus for the circumference, has the advantage of much greater convergency than that above.

(68.) From the foregoing integral expression for the area it appears that although the integral $\int \sqrt{r^2 - x^2}\, dx$ cannot be accurately expressed in finite terms, yet it may always be supplied by a circular area AB, the sine of the bounding arc being $\mathrm{OA} = x$; the sine of a similar arc to this, but of radius unity, instead of r is $\dfrac{x}{r}$; therefore, since similar parts of circles are as the squares of their radii, the above area will be equal to r^2 times a circular area to sine $\dfrac{x}{r}$, the radius of which is unity. If, therefore, a table of semi segments CBD were calculated for all values of DC, from $\mathrm{DC} = 0$ to $\mathrm{DC} = \mathrm{O\,E} = 1$, such a table would greatly facilitate the calculation of definite integrals of the above form, for it would then be merely necessary to add to the tabular number the rectangle D A, and to multiply by r^2. If the origin of the axis had been placed at the extremity F of the diameter, then the integral expressing the area would have been $\int \sqrt{2rx - x^2}\,.\, dx$, which, as above, may be expressed by a circular area of radius unity, so that

$$\int \sqrt{r^2 - x^2} \cdot dx = r^2 \times \frac{1}{2} \text{ circular zone, sine} = \frac{x}{r}$$

$$\int \sqrt{2rx - x^2} \cdot dx = r^2 \times \frac{1}{2} \text{ circular segment, ver. sin.} = \frac{x}{r}.$$

By reference to the figure it will be further obvious that these expressions are the same as

$$\int \sqrt{r^2 - x^2} \cdot dx = \frac{1}{2}\, r^2\, \sin.^{-1} \frac{x}{r} + \frac{1}{2}\, x\, \sqrt{r^2 - x^2}$$

$$\int \sqrt{2rx - x^2} \cdot dx = \frac{1}{2} r^2 \,\text{versin.}^{-1} \frac{x}{r} - \frac{1}{2} (r - x) \sqrt{2rx - x^2}$$

the first being the sector OBC plus the triangle OCA, and the second the sector OC′F minus the triangle OC′A′.

3. To determine the area of an ellipse

$$y = \frac{b}{a} \sqrt{a^2 - x^2}$$

$$\therefore \int y\, dx = \frac{b}{a} \int \sqrt{a^2 - x^2} \cdot dx\,;$$

but as we have just seen $\int \sqrt{a^2 - x^2} \cdot dx$ is the expression for a circular area whose radius is a, it follows, therefore, that if a circle circumscribe an ellipse, the area of the ellipse will be to that of the circle as the minor axis to the major; we have then only to multiply the area of the circumscribing circle by the minor diameter, and to divide the product by the major diameter, and we shall have the area of the ellipse.

4. To determine the area comprehended between the curve and asymptotes of an hyperbola.

The equation of an hyperbola between the asymptotes is (*Anal. Geom.* p. 120)

$$xy = \frac{a^2 + b^2}{4},$$

the axes ON, OK being inclined at an angle ϕ; hence

$$\text{sin.}\ \phi y\, dx = \frac{a^2 + b^2}{4}\ \text{sin.}\ \phi\ \frac{dx}{x}$$

$$\therefore u = \frac{a^2 + b^2}{4}\ \text{sin.}\ \phi \int \frac{dx}{x} = \frac{a^2 + b^2}{4}\ \text{sin.}\ \phi \log. x + C.$$

If the area be supposed to commence when $x = 1$ then $C = 0$, and the expression for the area included between one of the curvilinear and asymptotic legs, measured from this point, will be

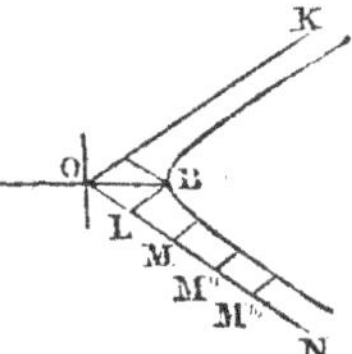

$$u = \frac{a^2 + b^2}{4}\ \text{sin.}\ \phi \log. x \ \ldots \ (1).$$

Let L be the point of commencement, that is, let OL represent unity,

then since the rhombus OB is

$$\mathrm{OB}^2 \sin. \phi = \frac{a^2 + b^2}{4} \sin. \phi,$$

the coefficient of log. x in (1) is simply sin. ϕ, because, by hypothesis, OB $= 1$; hence then the hyperbolic spaces BM, BM', &c. will be generally represented by

$$u = \sin. \phi \, \log. x.$$

so that, if OM, OM', &c. represent any series of numbers agreeably to the scale OL $= 1$, the spaces BM, BM', &c. will truly represent the logarithms of those numbers taken according to that system whose modulas is sin. ϕ, to radius OL; and thus, by varying the inclination of the asymptotes, innumerable systems of logarithms may be represented by hyperbolic spaces. If we wish to represent in this way Napier's system, called usually hyperbolic logarithms, although in strictness every system has equal claims to such a designation, we must make sin. $\phi = 1$, that is, the hyperbola must be equilateral. If we wish to represent Briggs's, or the common system, we must make sin. $\phi = \cdot 43429448^{*} \ldots \ldots, \therefore \phi = 25^{\circ}, 55', 16''$.

5. Let AFBF' be a given circle, and AB a diameter; let the radius OT revolve round the centre O, and let OP be always perpendicular to OT, meeting TP drawn parallel to AB in P. Required the equation and quadrature of the curve which is the locus of P.

Take the centre O for the origin, the fixed line OB for axis of x, and the perpendicular OF for axis of y. Put the radius OT $= a$; then, by similar triangles OPG, TOP, we have

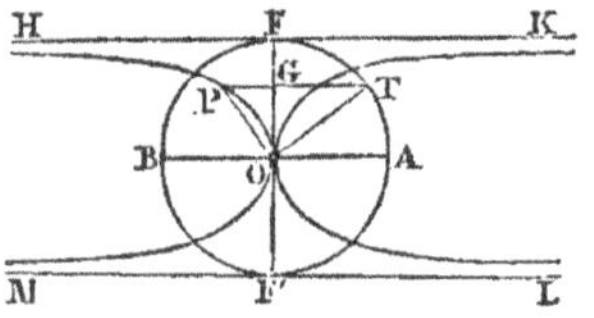

$$\mathrm{OG}^2 : \mathrm{PG}^2 :: \mathrm{OT}^2 : \mathrm{OP}^2,$$

that is,

$$y^2 : x^2 :: a^2 : x^2 + y^2$$

whence

$$a^2 x^2 = x^2 y^2 + y^4$$

---

* For the determination of this number, see my Essay on Logarithms, page 8.

$$\therefore x^2 = \frac{y^4}{a^2 - y^2} \quad \therefore x = \frac{y^2}{\sqrt{a^2 - y^2}},$$

the equation of the curve. When $x = 0$, $y = 0$, therefore the curve begins at O; when $y = \pm a$, $x = \infty$; hence the two tangents HK, ML are asymptotes to the curve.

For the quadrature we have

$$u = \int x \, dy = \int \frac{y^2 \, dy}{\sqrt{a^2 - y^2}},$$

this integral, by page 45, is

$$\tfrac{1}{2} a \cdot \text{arc AT} - \frac{1}{2} y \sqrt{a^2 - y^2} + C.$$

As the curve begins when $y = 0 \therefore C = 0$, and from this to $y = a$, between which a fourth part of the curve must be described, since the revolving radius will have passed through a quadrant, the area will be

$$\frac{1}{2} a \cdot \text{arc AF} = \text{quadrant AOF},$$

and therefore the whole area included by the four infinite branches of the curve, and the asymptotes HK, ML is equal to the area of the circle AFBF'.

6. Let ABC be a right-angled triangle, whose base AB is given, and in the variable hypothenuse produced, take CP, such that AC . CP may always be equal to BC^2. Required the quadrature of the curve which is the locus of P.

Let BX, BY be the axes of reference, and put $AB = a$, then we shall always have

$$AP = \sqrt{(a + x)^2 + y^2}$$

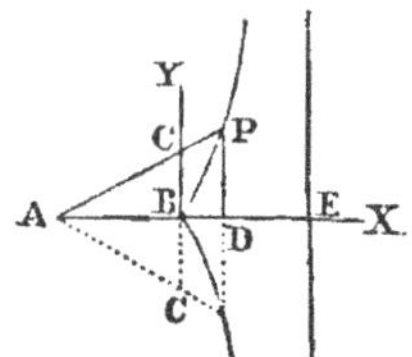

and, by similar triangles,

$$AD : AP :: AB : AC,$$

that is,

$$a + x : \sqrt{(a + x)^2 + y^2} :: a : \frac{a}{a + x} \sqrt{(a + x)^2 + y^2}$$

$$AD : AP :: BD : CP,$$

that is,

$$a + x : \sqrt{(a + x)^2 + y^2} :: x : \frac{x}{a + x} \sqrt{(a + x)^2 + y^2}$$

$$AD : DP :: AB : BC,$$

that is

$$a + x : y :: a : \frac{ay}{a + x} :$$

hence, by hypothesis,

$$\left(\frac{ay}{a + x}\right)^2 = \frac{ax}{(a + x)^2} \{(a + x)^2 + y^2\},$$

from which we get

$$y = \frac{ax + x^2}{\sqrt{ax - x^2}},$$

the equation of the curve. When $x = 0$, $y = 0$, and when x is negative, y is imaginary, $\therefore$ the curve begins at B; when $x = a$, $y = \infty$; hence, making $BE = BA$, the perpendicular through E will be an asymptote to the curve.

For the quadrature we have

$$\int y \, dx = \int \frac{ax \, dx}{\sqrt{ax - x^2}} + \int \frac{x^2 \, dx}{\sqrt{ax - x^2}}$$

$$= \frac{7}{8} a^2 \, \text{versin.}^{-1} \frac{2x}{a} - \left(\frac{7}{4} a + \frac{x}{2}\right) \sqrt{ax - x^2} \ (\text{see ex. 5, p. 48}),$$

the correction C is 0, because the integral ought to vanish for $x = 0$; hence, when $x = a$, we have

$$u = \frac{7}{8} a^2 \, \text{versin.}^{-1} 2 = \frac{7}{2} \cdot \frac{a}{2} \left(\frac{a}{2} \, \text{versin.}^{-1} 2\right),$$

the quantity within the parentheses is obviously the length of the semi-circle on BE; hence the area of the infinite space between the curve and asymptote is equal to 7 times the circle on BE.

7. ACB is a given semicircle, CD any ordinate: join AC, and draw

DP perpendicular to AC.   Required the quadrature of the curve which is the locus of P.

Draw CB, and put $AB = a$, $AD = z$, then

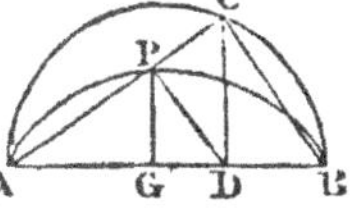

$$CD = \sqrt{az - z^2},$$

and (*Young's Geometry*, p. 91),

$$AB : AD :: \quad CD \quad : PG,$$

that is,

$$a : \quad z \quad :: \sqrt{az - z^2} : PG = \frac{z}{a} \sqrt{az - z^2} = y,$$

also

$$AB : AD :: \quad AD \quad : AG,$$

that is,

$$a : \quad z \quad :: \quad z \quad : AG = \frac{z^2}{a} = x,$$

consequently

$$\int y dx = \frac{2}{a^2} \int x^2 \, dx \sqrt{ax - x^2} = \frac{2}{a^2} \int x^{\frac{5}{2}} dx \sqrt{a - x}.$$

$$\int y \, dx = \frac{2}{a^2} \int x^2 \, dx \sqrt{ax - x^2} = \frac{2}{a^2} \int x^{\frac{5}{2}} \, dx \sqrt{a - x}.$$

This integral satisfies the condition of integrability at art. (21), and, if taken between the limits $x = 0$ and $x = a$, values for which y becomes 0, and beyond which y becomes imaginary, we shall have, for the area of the whole curve,

$$u = \frac{5}{8} \text{ semicircle ACB.}$$

8. To determine the area of the catenary.

In this curve we have found (ex. 10, p. 123,) that

$$s^2 = 2ax + x^2,$$

from which we get, by solving the quadratic,

$$x = \sqrt{a^2 + s^2} - a \therefore dx = \frac{sds}{\sqrt{a^2 - s^2}},$$

also, since universally

$$dy = \sqrt{ds^2 - dx^2}.$$

N

we have

$$dy = \sqrt{ds^2 - \frac{s^2\, ds^2}{a^2 + s^2}} = \frac{a\,ds}{\sqrt{a^2 + s^2}} \quad \ldots \ldots (1).$$

Multiplying this by the value of x, above, we have

$$x\,dy = a\,ds - \frac{a^2\, ds}{\sqrt{a^2 + s^2}} = a\,ds - a\,dy$$

$$\therefore u = \int x\,dy = a \int ds - a \int dy = as - ay \quad \ldots \ldots (2),$$

which requires no correction, since, s, y, and u vanish together. The area here found is that *below* the curve (*see fig. at* p. 123) included between the tangent through the vertex, the ordinate at the extremity of s, and s itself. Subtracting then this area from the rectangle

$$xy = y \sqrt{a^2 + s^2} - uy,$$

we have

$$u = y \sqrt{a^2 + s^2} - as,$$

or since, by integrating (1),

$$y = a \log. \frac{s + \sqrt{a^2 + s^2}}{a} \quad \ldots \ldots (3)$$

$$\therefore u = a \sqrt{a^2 + s^2} . \log. \frac{s + \sqrt{a^2 + s^2}}{a} - as \quad \ldots \ldots (4),$$

the area above the curve.

By putting in (3) for $\sqrt{a^2 + s^2}$ its value $a + x$, above, and for s its value given by the first equation, it becomes

$$y = a \log. \frac{a + x + \sqrt{2ax + x^2}}{a},$$

which is the equation of the catenary between x and y.

It is obvious that equation (3) may be written thus

$$ae^{\frac{y}{a}} = s + \sqrt{a^2 + s^2} \therefore ae^{\frac{y}{a}} - s = \sqrt{a^2 + s^2}$$

squaring each side, we have

$$-2se^{\frac{y}{a}} + ae^{\frac{2y}{a}} = a$$

$$\therefore s = \frac{1}{2}a\left\{e^{\frac{y}{a}} - e^{-\frac{y}{a}}\right\} \ldots \ldots (5).$$

Proceeding in like manner with equation (4), we have

$$x = \frac{1}{2}a\left\{e^{\frac{y}{a}} + e^{-\frac{y}{a}}\right\} - a \ldots \ldots (6).$$

9. To determine the quadrature of the *curve of sines* or *sinusoid*, its equation being $y = \sin. x$ to radius r

$$u = r\,(r - \sqrt{r^2 - y^2}.$$

10. To determine the area of the curve of tangents its equation being $y = \tan. x$ to radius r.

$$u = -\,r^2 \log. \cos. x.$$

11. To determine the area of the curve whose equation is

$$y = x^x,$$

between the limits $x = 0,\ x = 1$,

$$u = 1 - \frac{1}{2^2} + \frac{1}{3^3} - \frac{1}{4^4} + \frac{1}{5^5} - \&c.$$

12. To determine the area of the logarithmic curve, its equation being

$$y = a^x$$

$$u = \frac{y - 1}{\log. a}.$$

13. Prove that the area of the common cycloid is equal to three times the area of the generating circle.

14. ACB is a given semicircle, and DC any ordinate; bisect the arc AC in E, and join AC and AE, and upon DC take DF equal to the sum of the chords AC, AE. Required the area of the whole curve, which is the locus of F.

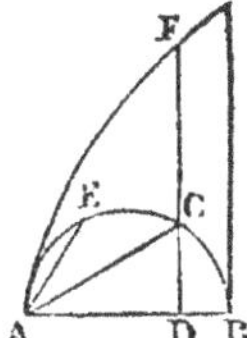

$$u = \frac{40 + 16\sqrt{2}}{15}\,a^2.$$

15. AEB is a straight line bisecting perpendicularly the given straight line DEC.   From D, one of the extremities of CD, draw the straight line DFP, cutting AB in F, and make FP equal to EF.   Required the area of the curve which p descri bes.

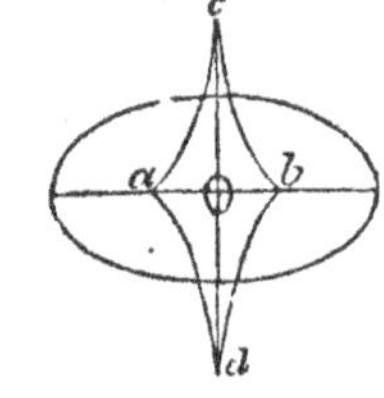

$$u = \text{semicircle on DC} + 2EC^2.$$

16. To determine the area $abcd$ included by the four branches of the evolute of the ellipse.

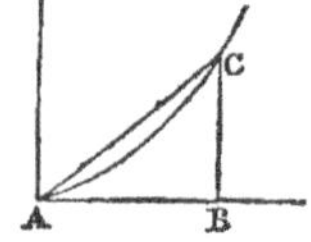

(69.) We shall now investigate the formula for the quadrature when the curve is referred to polar instead of rectangular coordinates.

In finding the lengths of curves, the formula of rectification was changed from rectangular to polar coordinates, by merely changing the independent variable from x to ω, the analytical value of the expression remaining unaltered; but in the formula for quadrature this change is not sufficient, for the analytical value of the function representing the area requires to be altered, since the spaces between the curve and the rectangular coordinates of any point in it, and the space between the curve and polar coordinates of the same point are themselves different.   Thus the space between thc curve AC, and the rectangular coordinates of the point C is ACB, but the space in reference to the polar coordinates of the same point is ACA, so that, calling, as before, the former space u, the latter will be $\frac{1}{2} xy - u$, it is therefore, the differential of this expression, taken relatively to x, that we are to transform to an equivalent differential, having ω for the independent variable, and not the differential $\frac{du}{dx}$.   Hence, putting v for the area ACA, and differentiating relatively to x, we have

$$\frac{dv}{dx} = \frac{1}{2}\left(x\,\frac{dy}{dx} + y\right) - \frac{du}{dx},$$

and changing the independent variable from x to ω, and multiplying by $\frac{dx}{d\omega}$, there results

$$\frac{dv}{d\omega} = \frac{1}{2}\left(x\,\frac{dy}{d\omega} + y\,\frac{dx}{d\omega}\right) - \frac{du}{dx}\,\frac{dx}{d\omega},$$

or, since $\dfrac{du}{x\omega} = y$,

$$\frac{dv}{d\omega} = \frac{1}{2}\left(x\,\frac{dy}{d\omega} - y\,\frac{dx}{d\omega}\right),$$

in which equation, if we substitute for $x, y, \dfrac{dx}{d\omega}$ and $\dfrac{dy}{d\omega}$, their values in terms of r and ω, as given at page 117 of the *Diff. Calc.* we shall have for the formula sought

$$\frac{dv}{d\omega} = \frac{1}{2}\,r^2 \therefore v = \frac{1}{2}\int r^2\,d\omega,$$

r being a function of ω, given by the polar equation of the curve.

EXAMPLES.

(70.)   1. To determine the area of the spiral of Archimedes.

$$r = a\omega \therefore \frac{1}{2}\int r^2\,d\omega = \frac{1}{2a}\int r^2\,dr = \frac{r^3}{6a} + C.$$

If the area is measured from the pole, then $u = 0$, when $r = 0 \therefore C = 0$ and the expression for the area is

$$v = \frac{r^3}{6a};$$

but, if the area is measured from the point $r = r'$, then v is 0, when $r = r' \therefore C = -\dfrac{r^3}{6a}$, and the expression is

$$v = \frac{r^3 - r'^3}{6a}$$

for the area comprehended between the two radii vectores r', r.

2. Tangents are drawn to an equilateral hyperbola, and perpendiculars to them are drawn from the centre: required the equation and quadrature of the curve, which is the locus of the intersections.

Taking the diameters of the hyperbola for axes, we have for any point (x', y') the equation

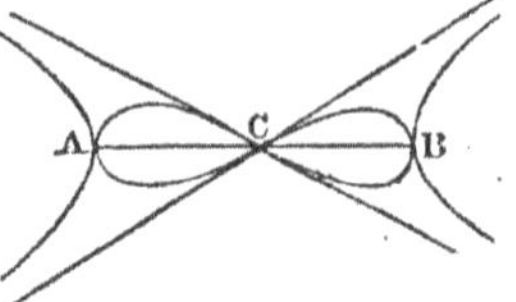

$$y'^2 + x'^2 = -a^2,$$

and for the equation of the tangent through it

$$y - y' = \frac{x'}{y'}(x - x'),$$

also for the perpendicular to this from the origin

$$y = -\frac{y'}{x'}\, x.$$

Eliminating x', y', by means of these three equations, we have between (x,y), the intersection of the two latter lines, the eqnation

$$a^2(y^2 - x^2) + (y^2 + x^2)^2 = 0$$

for the .ocus sought. This curve is called the *lemniscata of Bernouilli*.

To transform the above equation from rectangular to polar coordinates, we must substitute for x and y the values

$$x = r \cos.\,\omega\,;\ y = r \sin.\,\omega,$$

which will give the polar equation

$$r^2 - a^2 \cos.^2 \omega - \sin.^2 \omega) = 0,$$

or, substituting cos. 2ω for its equal cos.$^2 \omega$ — sin.$^2 \omega$,

$$r^2 - a^2 \cos.\,2\omega = 0,$$

from which it will be easy to discuss the figure of the curve. Thus, when $\omega = 0$, $r = \pm a$; hence the curve passes through the extremities of the transverse axis of the hyperbola; it also touches the hyperbola at those points, for if r were any where greater than a, cos. 2ω would, by the above equation, be greater than 1, which is impossible. When $r = 0$, ω must be either $\dfrac{\pi}{4}$, $\dfrac{3\pi}{4}$, $\dfrac{5\pi}{4}$ or $\dfrac{7\pi}{4}$, so that the curve passes through the centre of the hyperbola in four directions, being, indeed, the directions of the asymptotes; these are, therefore, tangents to it: hence the curve consists of two *leaves*, as represented in the diagram.

For the quadrature we have, by differentiating the polar equation,

$$r\, dr = -\, a^2 \sin. 2\omega\, d\omega$$

$$\therefore d\omega = -\, \frac{r dr}{a^2 \sin. 2\omega} = -\, \frac{r dr}{\sqrt{a^4 - r^4}}$$

$$\therefore \frac{1}{2} \int r^2\, d\omega = -\, \frac{1}{2} \int \frac{r^3\, dr}{\sqrt{a^4 - r^4}} = \frac{1}{4} \sqrt{a^4 - r^4} + C.$$

Between the limits $r = a$ and $r = 0$, which comprehends the upper segment BC, the area is $\frac{1}{4} a^2$, which, being a quarter of the whole area, we have for the entire curve

$$v = a^2,$$

the square described on the the semi axis of the hyperbola.

3. To determine the quadrature of the curve in example 7, page 133, by means of its polar equation.

Call the diameter AB, a, $\mathrm{AP} = r$, $\mathrm{PAB} = \omega$, then we have

$$\mathrm{AC} = \mathrm{AB} \cos. \omega, \quad \mathrm{AD} = \mathrm{AC} \cos. \omega, \quad \mathrm{AP} = \mathrm{AD} \cos. \omega$$

$$\therefore r = a \cos.^3 x$$

the polar equation of the curve

$$\therefore \frac{1}{2} \int r^2\, d\omega = \frac{a^2}{2} \int \cos.^6 \omega\, d\omega,$$

this integral is (34)

$$\frac{a^2}{2} \cdot \frac{\sin. x}{6} \left\{ \cos.^5 x + \frac{5}{4} \cos.^3 x + \frac{5 \cdot 3}{4 \cdot 2} \cos. x \right\} +$$

$$\frac{a^2}{2} \cdot \frac{5 \cdot 3 \cdot 1}{6 \cdot 4 \cdot 2} x + C,$$

which, between the limits $\omega = 0$, $\omega = \dfrac{\omega}{2}$, comprehending the whole curve, is

$$v = \frac{5}{8} \cdot \frac{a}{2} \cdot \frac{\pi}{2},$$

which is $\dfrac{5}{8}$, the semicircle ACB.

4. AM, AN are straight lines perpendicular to each other and a straight line BD of given length having its extremities always in them

is moved from a horizontal to a vertical position. If in BD, BF be taken always equal to BA, what will be the area of the curve in which F is always found.

Put $DAF = \omega$, $AF = r$, and $BD = a$, also draw BE perpendicular to AF, which will bisect the angle FBA of the isosceles triangle BFA, and the angle ABE will be equal to the angle DAF.

Now $AB = BD$ cos. $ABD = BD$ cos. 2ω; $AP = 2AB$ sin. $\omega = 2BD$ sin. ω cos. 2ω, that is,

$$r = 2a \text{ sin. } \omega \text{ cos. } 2\omega,$$

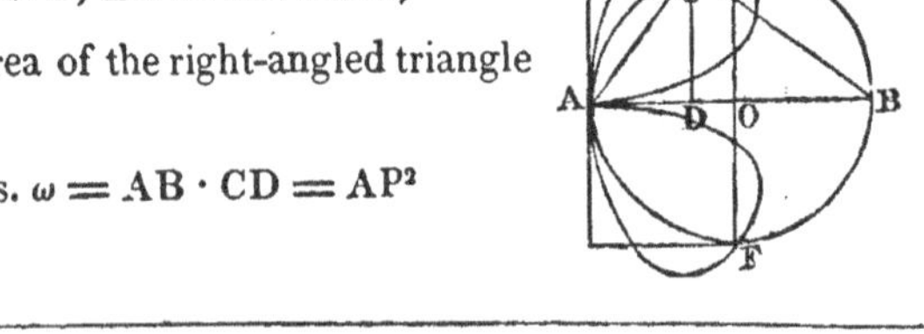

the polar equation of the curve; hence

$$\frac{1}{2}\int r^2 \, d\omega = 2a^2 \int \text{sin.}^2\omega \text{ cos.}^2 2\omega \, d\omega,$$

or, substituting (*Gregory's Trig.* p. 43,)

$$1 - 2 \text{ sin.}^2 \omega \text{ for cos. } 2\omega$$

$$\frac{1}{2}\int r^2 \, d\omega = 2a^2 \int \text{sin.}^2 \omega \, d\omega - 4\int \text{sin.}^4 \omega \, d\omega + 4\int \text{sin.}^6 \omega \, d\omega,$$

which for the whole curve, or between the limits $\omega = 0$, $\omega = 45°$ becomes[*]

$$x = \cdot 0594a^2.$$

5. AB is the diameter of a given circle, AC any chord, CD perpendicular to AB, and P a point in AC, so taken that $AP^2 = AB \cdot CD$. Required the quadrature of the curve which is the locus of P.

Put $AB = a$, and $PAB = \omega$, then

$$AC = AB \text{ cos. } \omega, \quad BC = AB \text{ sin. } \omega,$$

therefore twice the area of the right-angled triangle CAB is

$$AB^2 \text{ sin. } x \text{ cos. } \omega = AB \cdot CD = AP^2$$

that is,

---

[*] It is obvious, from a slight examination of the equation of the curve, that the entire locus of that equation consists of four *leaves,* symmetrically situated round the point A, but only one of these can come within the geometrical restrictions of the problem.

$$r^2 = a^2 \sin. x \cos. \omega = \frac{a^2}{2} \sin. 2\omega,$$

the polar equation, from which it appears that $r = 0$, both when $x = 0$, and when $\omega = 90°$, so that, while the radius vector passes through $90°$ on either side of AB, one *leaf* of the curve will be described.*

For the area of this leaf we have

$$\frac{1}{2} \int r^2 \, d\omega = \frac{a^2}{2} \int \sin. \omega \cos. \omega \, d\omega = -\frac{a^2}{4} \cos.^2 \omega + C,$$

which, between the limits of the curve is

$$v = \frac{a^2}{4} = \text{the square on AO,}$$

so that the whole area of the two leaves is equal to half the square of the diameter, or to the rectangle EF.

6. To determine the area included between two radii vectores r', r, of a logarithmic spiral, its equation being $r = a^{\omega}$.

$$v = \frac{m \, (r^2 - r')}{4},$$

where m is the modulus of the system of logarithms whose base is a.

7. To determine the area included between two radii vectores r', r, of a hyperbolic spiral, its equation being $r = \frac{a}{\omega}$.

$$v = \frac{a \, (r - r')}{2}.$$

8. To determine the polar equation and quadrature of the involute of the circle.

$$\text{Polar equation } a\omega = \sqrt{r^2 - a^2} - a \cos.^{-1} \frac{a}{r}$$

$$\text{Quadrature } v = \frac{(r^2 - a^2)^{\frac{3}{2}}}{6a}$$

---

* It is obvious that the curve can consist of but two *leaves* or *loops*, as well from the geometrical restrictions of the locus, as from its analytical representation above; for when ω exceeds $90°$, r becomes imaginary

9. From a given point in the circumference of a given circle any chord is drawn, and from its extremity a line is drawn to the centre. If from the centre a perpendicular to this line be drawn, what will be quadrature of the curve which is the locus of its intersection with the chord?

The whole area of the locus is $\cdot 4292 \times rad^2$ of circle.

10. Between the sides of a right angle a straight line is drawn so as to enclose a given area; if from the vertex of the right angle a perpendicular to this line be drawn, what will be the quadrature of the curve, which is the locus of the intersection.

The area is half that of the triangle.

CHAPTER III.

ON THE QUADRATURE OF CURVE SURFACES, AND ON THE CUBATURE OF VOLUMES.

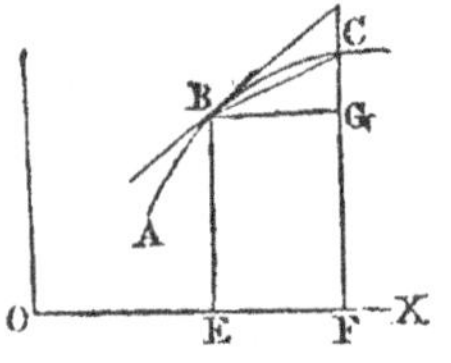

(71.) Let AB be an arc of any plane curve, and let it be required to find the differential expression for the surface generated by the revolution of this arc about one of the coordinate axes, as the axis of x.

Let BC be any increment of the arc AB, and draw the chord BC, and the tangent BD, meeting the ordinate FC in D. Then, since the bent line BDC envelopes the curve line BC, if the system of lines revolve round AX, the surface generated by BCD will envelope that generated by the curve BC; the former surface, therefore, will be greater than the latter (*See Note* C). Again, since the curve BC envelopes the line BC, the surface generated by the revolution of the former will be greater than that generated by the revolution of the latter. Hence the surface generated by the arc BC is always of intermediate magnitude between the surfaces generated by BDC and by BC. Let us then seek the limit of the ratio of the two latter surfaces, for, if this should be unity, then the ratio of the intermediate surface to either must in the limit be necessarily unity.

The surface generated by BD is a conical frustum or trunk, and we know, by the principles of common geometry, that every such surface is equal to its side, or the generating line multiplied by half the sum of the circumferences of the two ends; also the annular space generated by DC is equal to its length multiplied by half the sum of the circumferences generated by the ends D, C.

Hence, for the largest of our three surfaces we have the expression

$$\pi \mathrm{BD}\,(\mathrm{EB} + \mathrm{FD}) + \pi \mathrm{CD}\,(\mathrm{FC} + \mathrm{FD}) \;.\;.\;.\;.\; (1),$$

and for the smallest the expression

$$\pi \mathrm{BC}\,(\mathrm{EB} + \mathrm{FC}) \;.\;.\;.\;.\; (2),$$

and we have now to ascertain the ratio of these expressions in the limit, that is, when $\mathrm{EF} = h = 0$. To effect this more readily it will be requisite to develope these expressions according to the powers of h. Now it is obvious that

$$\mathrm{BD} = \sqrt{h^2 + \frac{dy^2}{dx^2}\,h^2}, \quad \mathrm{EB} + \mathrm{FD} = 2y + \frac{dy}{dx}\,h,$$

$$\mathrm{CD} = y + \frac{dy}{dx}\,h - \left\{ y + \frac{dy}{dx}\,h + \frac{d^2y}{dx^2}\cdot\frac{h^2}{1\cdot 2} + \&c. \right\},$$

$$\mathrm{FC} + \mathrm{FD} = 2y + 2\frac{dy}{dx}\,h + \frac{d^2y}{dx^2}\cdot\frac{h^2}{1\cdot 2} + \&c.$$

also

$$\mathrm{BC} = \sqrt{\mathrm{BG^2 + GC^2}} = \sqrt{\left\{ h^2 + \left(\frac{dy}{dx}\,h + \frac{d^2y}{dx^2}\,\frac{h^2}{1\cdot 2} + \&c.\right)^2\right\}}$$

$$\mathrm{EB} + \mathrm{FC} = 2y + \frac{dy}{dx}\,h + \frac{d^2y}{dx^2}\,\frac{h^2}{1\cdot 2} + \&c.$$

hence, by substitution, the expression (1) is

$$\pi\,h\,\sqrt{\left(1 + \frac{dy^2}{dx^2}\right)}\left(2y + \frac{dy}{dx}\,h\right) - \pi\left(\frac{d^2y}{dx^2}\cdot\frac{h^2}{1\cdot 2} + \&c.\right)\left(2y + \right.$$

$$\left. 2\frac{dy}{dx}\,h + \&c.\right)$$

and the expression (2)

$$\pi h \sqrt{\{1 + (\frac{dy}{dx} + \frac{d^2y}{dx^2}\ \frac{h}{2} + \&c.)^2\}}\ \{2y + \frac{dy}{dx} h +$$

$$\frac{d^2y}{dx^2}\ \frac{h^2}{1 \cdot 2} + \&c.\}$$

Dividing each of these by h, and then putting $h = 0$, their limiting ratio is

$$\frac{\pi \sqrt{1 + \frac{dy^2}{dx^2}} \cdot 2y}{\pi \sqrt{1 + \frac{dy^2}{dx^2}} \cdot 2y} = 1,$$

consequently, when $h = 0$, then

$$\frac{\text{inc. of surface}}{h} = \frac{dS}{dx} = 2\pi y \sqrt{1 + \frac{dy^2}{dx^2}}$$

$$\therefore\ S = 2\pi \int y \sqrt{1 + \frac{dy^2}{dx^2}}\ dx = 2\pi \int yds,$$

$$\text{or}\ S = 2\pi \int y \sqrt{1 + \frac{dx^2}{dy^2}} \cdot dy.^*$$

We shall now apply this general formula to some examples:

EXAMPLES.

(72.)   1 To determine the surface of a sphere.
The equation of the generating circle is

$$x^2 + y^2 = r^2$$

---

* To persons familiar with other modes of investigation the above process may seem unnecessarily long; but it is apprehended that most of the shorter methods will be found, upon examination, to be deficient in rigour.   Thus *Francœur* (*Cours de Math. tom.* 2, p. 341,) reasons as if the surface generated by the arc were of intermediate magnitude between those generated by the chord and tangent, which is not necessarily the case.

$$\therefore\ x\,dx + y\,dy = 0 \ \therefore \frac{dy^2}{dx^2} = \frac{x^2}{y^2};$$

consequently

$$S = 2\pi \int \sqrt{1 + \frac{dy^2}{dx^2}}\ dx = 2\pi \int \sqrt{x^2 + y^2}\ dx$$

$$= 2\pi \int r\,dx = 2\pi r x + C.$$

Between the limits $x = 0$, $x = r$, this is $2\pi r^2$; hence the surface of the whole sphere is

$$s = 4\pi r^2,$$

which is equal to four times the area of one of its great circles.

2. To determine the surface of a spheroid.

The equation of the generating ellipse is

$$a^2 y^2 + b^2 x^2 = a^2 b^2,$$

or

$$y = \sqrt{(1 - e^2)(a^2 - x^2)},$$

from which we have already found (*art.* 61)

$$ds = \frac{\sqrt{a^2 - e^2 x^2}}{\sqrt{a^2 - x^2}}\ dx,$$

consequently

$$S = 2\pi \int y\,ds = 2\pi\ \frac{e}{\sqrt{1 - e^2}} \int \sqrt{\frac{a^2}{e^2} - x^2}\ dx,$$

but $\sqrt{1 - e^2} = \dfrac{b}{a}$, therefore

$$S = \frac{2\pi\,be}{a} \int \sqrt{\frac{a^2}{e^2} - x^2}\ dx.$$

Now, as we have observed at (68), the integral

$$\int \sqrt{\frac{a^2}{e^2} - x^2} \cdot dx$$

is equal to half a circular zone of radius $\dfrac{a}{e}$, the sine of the arc being x; therefore, calling this circular area A, we have

$$S = \frac{2\pi\, be}{a}\, A.$$

If $x = a$, then this expression will represent half the entire surface, or, calling the corresponding value of A, A′, we have for the whole surface of the spheroid

$$S = \frac{4\pi be}{a}\, A'.$$

3. To determine the surface of a paraboloid of revolution.

The equation of the generating curve being

$$y^2 = px,$$

we have

$$2y\, dy = p\, dx \therefore \frac{dx^2}{dy^2} = \frac{4y^2}{p^2}$$

hence

$$S = 2\pi \int y\, \sqrt{1 + \frac{dx^2}{dy^2}} \cdot dy = \frac{2\pi}{p} \int \sqrt{4y^2 + p^2} \cdot y\, dy.$$

But

$$\int (4y^2 + p^2)^{\frac{1}{2}}\, y\, dy = \frac{1}{12}\, (4y^2 + p^2)^{\frac{3}{2}} + C.$$

$$\therefore S = \frac{\pi}{6p}\, (4y^2 + p^2)^{\frac{3}{2}} + C,$$

which, between the limits $y = 0$, $y = y$, is

$$S = \frac{\pi}{6p}\, \{(4y^2 + p^2)^{\frac{3}{2}} - p^3\}.$$

4. To determine the surface generated by the revolution of a catenary about its axis.

We have already found (p. 123) that in this curve

$$s^2 = 2ax + x^2$$

$$\therefore a^2 + s^2 = a^2 + 2ax + x^2 \therefore \sqrt{a^2 + s^2} = a + x;$$

hence

$$dx = \frac{s\, ds}{\sqrt{a^2 + s^2}},$$

consequently

$$dy = \sqrt{ds^2 - dx^2} = \frac{a\,ds}{\sqrt{a^2 + s^2}}$$

Now

hence
$$\int y\,ds = ys - \int s\,dy;$$

$$S = 2\pi ys - 2a\pi \int \frac{s\,ds}{\sqrt{a^2 + s^2}}$$

$$= 2\pi ys - 2a\pi \sqrt{a^2 + s^2} + C.$$

When $s = 0$ then $S = 0 \therefore 0 = -2a^2\pi + C \therefore C = 2a^2\pi$ therefore

$$S = 2\pi ys - 2a\pi \sqrt{a^2 + s^2} + 2a^2\,\pi,$$

which is the area of the surface, in terms of y and s, s being the length of the revolving arc measured from the vertex, and y the ordinate of its extremity. If for $\sqrt{a^2 + s^2}$ we substitute its value above, viz. $a + x$, the expression for the surface becomes, in terms of x, y and s,

$$S = 2\pi ys - 2a\pi x.$$

5. To determine the surface generated by the revolution of a cycloid about its base $2a$

$$S = \frac{16\pi}{3} a^2.$$

To determine the surface generated by a cycloid revolving round its axis, the diameter of the generating circle being a,

$$S = \left(\pi - \frac{4}{3}\right) 2\pi a^2.$$

Cubature.

(72.) Let us now investigate the differential expression for the volume bounded by the curve surface generated by the revolution of AB, and by the planes generated by the revolution of the ordinates of A and B. Let BC be an increment of the arc AB, taking care, however, that this increment be not so large that the ordinates between E and F may first increase, and then decrease, or that they may first decrease and then increase; but this interval must be taken so small that the ordinates from E to F may continually increase or continually diminish.

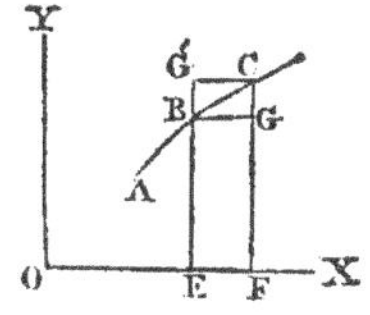

The necessity of this condition will appear, when we state that we are about to found our reasoning on this principle, viz. that the volume generated by BC is always intermediate in magnitude between the cylinders generated by BG, CG′, however we diminish the original increment $EF = h$, a principle which would not be necessarily true if either of the parallels BG, CG′, could cross the curve BC, and, to render this crossing impossible, the foregoing condition must obviously be observed: we say the solid generated by BC and the cylinder generated by BG, &c. for shortness, of course it is to be understood that the ordinates of the extremities of these lines revolve with them.

Admitting then that the solid increment generated by BC is always intermediate between the two cylinders, let us seek the limiting ratio of these latter.

The volume of a cylinder is equal to its base multiplied by its height, therefore the volume of the cylinder generated by BG is $\pi y^2 h$, and that generated by CG′ is

$$\pi \left(y^2 + \frac{dy^2}{dx} \, h + \frac{d^2 y^2}{dx^2} \; \frac{h^2}{1 \cdot 2} + \&c. \right) h,$$

the ratio of these is

$$\frac{\pi y^2}{\pi \left(y^2 + \dfrac{dy^2}{dx} \, h + \&c. \right)}$$

which when $h = 0$ becomes unity. Hence the ratio of the intermediate volume or the increment of the proposed volume to either of the cylinders is unity in the limit, that is

$$\frac{dV}{dx} = \pi y^2 \; \therefore \; V = \pi \int y^2 \, dx.$$

EXAMPLES.

(78.)   1. To determine the volume of a prolate spheroid.
Since the equation of the generating ellipse is

$$a^2 y^2 + b^2 x^2 = a^2 b^2$$

$$\therefore y^2 = \frac{b^2}{a^2} (a^2 - x^2)$$

$$\therefore V = \pi \int y^2 \, dx = \frac{\pi b^2}{a^2} \int (a^2 - x^2) \, dx = \pi b^2 \left(x - \frac{x^3}{3a^2} \right) + C,$$

and for the whole solid, that is, between the limits $x = a$ and $x = -a$, this integral is

$$V = \pi b^2 \left(2a - \frac{2}{3} a\right) = \frac{4}{3} \pi a b^2,$$

and since the circumscribing cylinder is $b\pi . b . 2a$, it follows that the volume of the spheroid is two thirds of its circumscribing cylinder.

When $a = b$ the spheroid becomes a sphere, which is therefore equal in volume to two thirds of its circumscribing cylinder.

2. To determine the volume of a paraboloid.

Since here $y^2 = px$ $\therefore$ $\pi \int y^2 dx = \pi p \int x \, dx = \frac{\pi p}{2} x^2 + C,$

$$\therefore V = \frac{\pi p}{2} x^2 = \frac{\pi}{2} y^2 x,$$

which is half the volume of the circumscribing cylinder.

3. To determine the volume generated by the revolution of the catenary about its axis.

By example 4, p. 146,

$$dx = \frac{s\,ds}{\sqrt{a^2 + s^2}}, \, dy = \frac{a\,ds}{\sqrt{a^2 + s^2}} = \frac{a\,dx}{s},$$

and the formula for cubature being decomposed is

$$\pi \int y^2 dx = \pi \left(y^2 x - 2\int xy \, dy\right) \, . \, . \, . \, . \, (1).$$

If for dy we put its value above, we have

$$xy \, dy = ay \, \frac{x\,dx}{s},$$

but from the equation of the curve

$$\frac{x\,dx}{s} = ds - \frac{a\,dx}{s} = ds - dy,$$

so that

$$xy \, dy = ay \, ds - ay \, dy,$$

and consequently

$$\int xy \, dy = a \int y\,ds - \frac{ay^2}{2} \, . \, . \, . \, . \, (2),$$

substituting this in (1) we have

$$V = \pi y^2 x + a \pi y^2 - 2 a \pi \int y \, ds,$$

but

$$2 \pi \int y \, ds = \text{Surface} = 2 \pi y s - 2 a \pi x \, ;$$

hence, finally,

$$V = \pi y^2 x + a \pi y^2 - 2 a \pi y s - 2 a^2 \pi x,$$

which requires no correction.

4. To determine the volume generated by the revolution of the parabola $y^n = ax$ about the axis of x:

$$V = \frac{n}{n+2} \cdot \pi y^2 x.$$

5. To determine the volume of a *parabolic spindle* which is generated by the revolution of a parabola about its base b, the height being a:

$$V = \frac{16 \pi a^2 b}{15}.$$

(74.) It ought to be remarked here, that there is another method in which volumes of revolution may be easily conceived to be generated, viz. by the motion of a curve parallel to its own plane and varying in magnitude according to a fixed law. Thus the volume of revolution upon which our observations in art. (72) are made, may be considered as generated by the motion of a circle whose radius is y; the centre being always on OX, its plane being always perpendicular to this line, and its radius varying according to the law of the variation of the ordinates y of the directrix ABC: and, viewing the generation as effected in this way, we may say, agreeably to the general result obtained in the article referred to, that *the differential of the volume is equal to the area of the generating circle multiplied by the differential of the axis.* This theorem is indeed true whatever be the generating area, provided only that, as in the case of the circle, it varies agreeably to some law dependent on the equation of the directrix; or, in other words, provided we can always express this in general terms as an invariable function of x and y, the general coordinates of the directrix, and therefore of x simply .

Thus, let us suppose AB in art. (72) to be the directrix which governs the magnitude of the generating surface whose edge is BE;

then, taking the same increment EC of the volume as before, we shall
have to constitute on the sections BE, FC instead of cylinders, prisms
BF, FG′, between which the increment of the volume will as before be
always intermediate. Now the ultimate ratio of these prisms is as
was shewn of the cylinders unity, for their volumes like those of the
cylinders are expressed by the product of their bases and heights: so
that, as we may by hypothesis represent the one base by fx, the other
will be $f(x + h)$, and therefore the ratio of the prisms themselves
will be

$$\frac{h f x}{h \left(fx + \dfrac{d f x}{dx} h + \dfrac{d^2 f x}{dx^2} \cdot \dfrac{h^2}{1 \cdot 2} + \&c. \right),}$$

which, by first dividing by h and then taking the limit or putting $h = 0$,
becomes unity as in the former case: the theorem above is therefore
generally true. Let us give an example of its application to the solid
called a *circular groin*. The generating area in this case is a square,
and the directrix, to which it is always perpendicular, a semicircle
passing through the middle points of two opposite sides.

Taking as axis of x the diameter perpendicular to the moving plane,
and the vertex as the origin, we have for any ordinate of the directrix,
that is, for half the side of the variable square, the expression

$$y = \sqrt{2ax - x^2};$$

hence, the area of the square is

$$4 (2ax - x^2),$$

therefore this multiplied by dx gives

$$dV = 4 (2ax - x^2)\, dx \therefore V = 4\int (2ax - x^2)\, dx$$

$$= 4ax^2 - \frac{4}{3} x^3,$$

C being 0 because the expression for V ought obviously to vanish
with x. As another example, let it be required to determine the volume
of an ellipsoid.

The equation of the ellipsoid being

$$a^2 b^2 z^2 + b^2 c^2 x^2 + a^2 c^2 y^2 = a^2 b^2 c^2,$$

it follows that the general equation of a section at any distance z from, and parallel to, the plane of xy, will be

$$a^2 y^2 + b^2 x^2 = \frac{a^2 b^2 (c^2 - z^2)}{c^2} ;$$

considering, therefore, this as the generating ellipse, we have for its semi-axes a', b',

$$a' = \frac{a \sqrt{c^2 - z^2}}{c}$$

$$b' = \frac{b \sqrt{c^2 - z^2}}{c}$$

and consequently for its area (see ex. 3, p. 129,)

$$\pi a' b' = \frac{\pi ab (c^2 - z^2)}{c^2} ;$$

hence, multiplying this by dz and integrating, we have

$$V = \frac{\pi ab}{c^2} \int (c^2 - z^2) \, dz$$

$$= \frac{\pi ab}{c^2} \left(c^2 z - \frac{1}{3} z^3 \right) + C.$$

This for the whole volume, or between the limits $z = c$, $z = -c$, becomes

$$V = \frac{4}{3} \pi abc,$$

which is equal to a sphere whose radius is equal to $\sqrt[3]{abc}$.

The formulas given in the present chapter for the quadrature of curve surfaces and the cubature of their volumes, will be found sufficient for most practical inquiries, as the curve surfaces presented to us in nature or employed in the arts are almost invariably surfaces of revolution. In order however to complete the subject of this chapter, we shall now investigate general expressions for the volume and surface of any body that can be represented by an equation.*

---

* The student may if he please pass over the remainder of this chapter, for the present. .

Let the equation of any curve surface be

$$z = \mathrm{F}(x, y).$$

Draw four planes parallel two and two to those of xz and yz, and let us seek the analytical expressions for the volume V and the surface S of the body MNEF, contained between these limits.

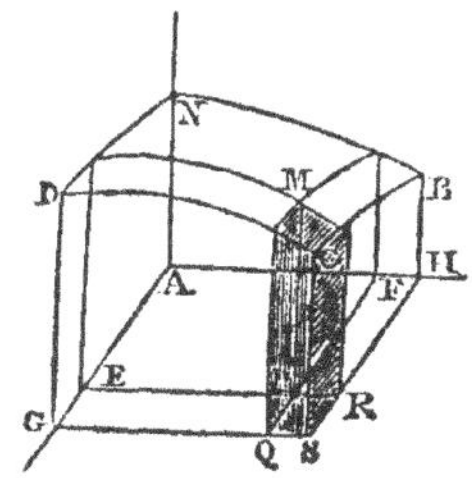

Let x and y take the increments h and k, by which means the point M or $(x, y, z,)$ will be removed to C, and the whole increment, taken by the body in consequence of this change, will consist of the two slices enclosed between the planes ME, SD; SB, FM. The sum of these expressed analytically, that is the increment $\mathrm{V}' - \mathrm{V}$, is (*Diff. Calc.* p. 84,)

$$\frac{d\mathrm{V}}{dx}\, h + \frac{d\mathrm{V}}{dy}\, k + \frac{d^2\mathrm{V}}{dx^2}\, \frac{h^2}{2} + \frac{d^2\mathrm{V}}{dx\,dy}\, hk + \frac{d^2\mathrm{V}}{dy^2}\, \frac{k^2}{2} + \&\mathrm{c}.$$

If from this total increment we take the two parts MG, MH, there will remain the column MS; the parts to be subtracted are

$$\mathrm{MG} = \frac{d\mathrm{V}}{dx}\, h + \frac{d^2\mathrm{V}}{dx^2}\, \frac{h^2}{2} + \&\mathrm{c}.$$

$$\mathrm{MH} = \frac{d\mathrm{V}}{dy}\, k + \frac{d^2\mathrm{V}}{dy^2}\, \frac{k^2}{2} + \&\mathrm{c}.$$

therefore the remainder is

$$\mathrm{MS} = \frac{d^2\mathrm{V}}{dx\,dy}\, hk + \&\mathrm{c}.$$

By the same process we find for the surface MC,

$$\mathrm{MC} = \frac{d^2\mathrm{U}}{dx\,dy}\, hk + \&\mathrm{c}. \qquad \frac{d^2S}{dx\,dy} hk$$

Now the volume MS is obviously always intermediate between two prisms on the same base, and of which the altitudes are respectively SC, PM.* As these prisms are to each other as their altitudes, we

---

* The increments h, k, are to be taken so small that the surface MC may be entirely convex or concave.

have for their ratio

$$\frac{z}{z + l},$$

l being the increment of z corresponding to the increments h and k; hence in the limit, or when $l = 0$, the ratio is unity, and consequently the ratio of the intermediate volume to either prism is unity in the limit, that is

$$\frac{\frac{d^2 V}{dx\,dy}\,hk}{zhk} = 1 \;\therefore\; \frac{d^2 V}{dx\,dy} = z,$$

therefore, multiplying by dy and integrating with respect to y, we have

$$\frac{dV}{dx} = \int z\,dy,$$

and multiplying by dx, and integrating with respect to x, we obtain finally

$$V = \int\int z\,dy\,dx,$$

so that the volume is obtained by a double integration, which is performed by integrating first on the supposition that one of the variables x, y, is constant, and then multiplying the integral properly corrected, or between the proper limits furnished by the problem, by the differential of the variable, at first considered constant, and integrating again. This will be best explained by an example.

Let the sphere be proposed, then

$$x^2 + y^2 + z^2 = r^2 \;\therefore\; z = \sqrt{r^2 - x^2 - y^2};$$

hence

$$\frac{d^2 V}{dx\,dy} = \sqrt{r^2 - x^2 - y^2},$$

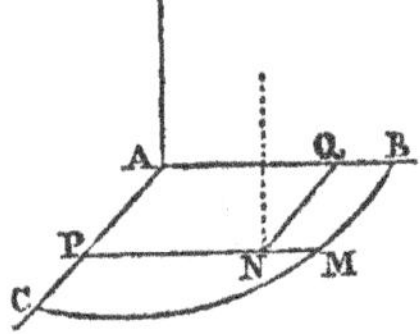

let us first integrate relatively to x, that is on the supposition that this is the only variable, and we shall have

$$\frac{dV}{dy} = \int \sqrt{r^2 - x^2 - y^2} \cdot dx,$$

in which integral y isconsidered a constant, therefore its value is (68) half a circular zone whose radius is $\sqrt{r^2 - y^2}$ and abscissa x.

It may be observed here, that the constant y being represented by AP, and the variable x by AQ or PN, this zone will obviously be that portion of the vertical section of the spherical quadrant which stands upon PN; the utmost limit to which x or PN can extend is from P to M, therefore, taking the foregoing integral between the limits $x = 0$,

$x = \mathrm{PM} = \sqrt{r^2 - y^2}$, it becomes $\dfrac{\pi}{4} (r^2 - y^2)$, which must therefore

express the area of the whole quadrant standing on PM. y is now to be considered as variable, and therefore this is a general expression for every vertical section of the spherical quadrant, of which the plane is parallel to AB; we know therefore from last article, as well as from the general expression for V in this, that

$$V = \frac{\pi}{4} \int (r^2 - y^2)\, dy = \frac{\pi}{4} \left(r^2 y - \frac{y^3}{3}\right),$$

which, between the limits $y = 0$, $y = r$, gives for the spherical quadrant

$$V = \frac{\pi r^3}{6},$$

and therefore the volume of the whole sphere is eight times this, or $\dfrac{4}{3} \pi r^3$.

(75.) Let us now consider the surface MC, for which we have already found the expression

$$MC = \frac{d^2 S}{dx\, dy}\, hk + \&c.$$

for the purpose of determining between what two surfaces, ultimately in a ratio of equality, it must be always intermediate. Let the sides of the vertical column MS be produced upwards, and still considering M to be the lowest, and C the highest point on the surface MC, conceive tangent planes to be drawn to the surface at M and C; of these the portions contained within the vertical planes will be the former greater in surface than the convex surface MC, and the latter less in surface, or the contrary if the surface MC is concave: it will be necessary to prove this.

And in order to this we may first observe that if, through the point of contact of one of these tangent parallelograms, any vertical plane be drawn, the linear section on the parallelogram will be a tangent to the curve section on the surface MC; and it is at once obvious that if, whatever be the direction of this section, the tangent is always greater or always less than the arc, both being terminated by the vertical sides of the column MS, the surface which is the locus of these tangents will accordingly be greater or less than the surface which is the locus of the arcs, that is than the surface MC: this being admitted, which is indeed axiomatical, our proof will be reduced to the shewing, that in the sections of which we have been speaking, the tangent really is in the one case always greater than the corresponding arc, and in the other case always less, and this is proved as follows.

Let MC be any arc, AM and BC the ordinates of its extremities, the former being the shorter and all the ordinates from A to B being in increasing order; then the tangent CM′ is obviously nearer to the perpendicular CH on AH than the chord CM, therefore CM′ is shorter than CM, and consequently

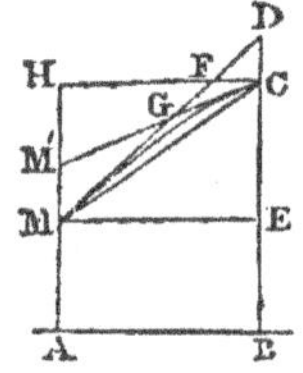

shorter than the arc MC. Also the tangent MD must be farther from the perpendicular ME than MC, and must therefore cross CH somewhere in F between the parallels. Now FC, FM, are together longer than the arc MC, (*Geom.* p. 204,) and FD is longer than FC, cousequently MD is longer than the arc MC. If the arc had been concave to AB instead of convex, we should have found by the same reasoning that CM′ would have been greater and MD less than the arc. Our proposition is therefore established, and it remains to find analytical expressions for the two tangent parallelograms between which we have now shown the surface MC must be always intermediate. We know that the tangent plane at the point (x, y, z,) or M, is inclined to the plane of xy at an angle of which the cosine is (*Diff. Calc.* p. 164,)

$$\frac{1}{\sqrt{1 + \dfrac{dz^2}{dx^2} + \dfrac{dz^2}{dy^2}}},$$

and as a parallelogram situated in space is equal to its projection on the plane of xy divided by the cosine of its inclination to that plane, it follows that since the projection of either of our parallelograms is

$= hk$, we must have for the area of tha ttouching at M the expression

$$hk \sqrt{1 + \frac{dz^2}{dx^2} + \frac{dz^2}{dy^2}};$$

the quantity under the radical being a function of x and y we may represent it by $f(x, y)$, and to find the expression for the other parallelogram or that touching at C, we shall merely have to change in this x into $x + h$ and y into $y + k$, multiplying the result as before by hk. Now

$$f(x + h, y + k) = f(x, y) + \frac{df(x, y)}{dx} h + \frac{df(x, y)}{dy} k + \&c.$$

consequently, the general expression for the ratio of the two parallelograms in space is

$$\frac{\sqrt{1 + \frac{dz^2}{dx^2} + \frac{dz^2}{dy^2}}}{\sqrt{1 + \frac{dz^2}{dx^2} + \frac{dz^2}{dy^2}} + \&c.}$$

which when h and k become 0 reduces to unity; hence in the limit the ratio of the intermediate surface MC to either of these parallelograms is unity, so that then

$$\frac{\frac{d^2 S}{dx\, dy} + \&c.}{\sqrt{1 + \frac{dz^2}{dx^2} + \frac{dz^2}{dy^2}}} \quad \text{becomes} \quad \frac{\frac{d^2 S}{dx\, dy}}{\sqrt{1 + \frac{dz^2}{dx^2} + \frac{dz^2}{dy^2}}} = 1$$

therefore

$$\frac{d^2 S}{dx\, dy} = \sqrt{1 + \frac{dz^2}{dx^2} + \frac{dz^2}{dy^2}} \quad \therefore \quad S = \iint \sqrt{1 + \frac{dz^2}{dx^2} + \frac{dz^2}{dy^2}} \cdot dx\, dy.$$

As an illustrative example let us take as before a spherical surface. From its equation

$$x^2 + y^2 + z^2 = r^2,$$

we get by differentiating

$$\frac{dz}{dx} = - \frac{x}{z}, \quad \frac{dz}{dy} = - \frac{y}{z},$$

and these values substituted in the foregoing expression under the radical reduces it to

$$\sqrt{1 + \frac{x^2}{z^2} + \frac{y^2}{z^2}} = \frac{1}{z}\sqrt{x^2 + y^2 + z^2} = \frac{r}{z}$$

$$= \frac{r}{\sqrt{r^2 - x^2 - y^2}};$$

hence

$$\frac{d^2 S}{dx\,dy} = \frac{r}{\sqrt{r^2 - x^2 - y^2}}.$$

Integrating relatively to x, y being, for the present, considered constant, we have

$$\frac{dS}{dy} = r \int \frac{dx}{\sqrt{r^2 - y^2 - x^2}} = r \sin.^{-1} \frac{x}{\sqrt{r^2 - y^2}}:$$

this taken as before between the limits P and M, (see fig. p. 154,) or from $x = 0$ to $x = \sqrt{r^2 - y^2}$ is $\dfrac{\pi r}{2}$, which is the quadrantal vertical arc subtended by PM; and multiplying by dy, and integrating again, have finally

$$S = \frac{\pi r}{2} y,$$

which from $y = 0$ to $y = r$ becomes for a quarter of the hemisphere $\dfrac{\pi r^2}{2}$, and therefore the surface of the whole sphere is $4\pi r^2$.

CHAPTER· III.

MISCELLANEOUS INTEGRATIONS.

(76.) The present may be considered as a supplementary chapter to what has already been said on the integration of functions of one variable, and in which it is intended to exhibit a few examples of the transformations to be effected in order to bring certain differential expressions to integrable forms.

EXAMPLES.

1. Required the integral of

$$\frac{\sqrt{2ax + x^2}}{x}\, dx.$$

Multiplying both numerator and denominator of the function by the numerator it becomes

$$\frac{(2a + x)\, dx}{\sqrt{2ax + x^2}} = \frac{adx + xdx}{\sqrt{2ax + x^2}} + \frac{adx}{\sqrt{2ax + x^2}}.$$

The proposed differential is thus divided into two others immediately integrable: the first by the rule for powers art. (4), and the second is integrated at p. 34, ex. 12; hence

$$\int \frac{\sqrt{2ax + x^2}}{x}\, dx = \sqrt{2ax + x^2} + a \log. (x + a + \sqrt{2ax + x^2}) + C.$$

It may be proper to observe that as the proposed satisfies the criterion of integrability at (21), it may be integrated by the method of substitution there explained.

2. Required the integral of

$$\frac{x\, dx}{\sqrt{ax + x^2}}.$$

By first adding $\frac{1}{2} a\,dx$ to, and then subtracting it from, the numerator, we shall convert this differential into two others, of which one will obviously be immediately integrable by the rule for powers; thus we shall have to integrate

$$\frac{\frac{1}{2} a\,dx + x\,dx}{\sqrt{ax + x^2}} - \frac{\frac{1}{2} a\,dx}{\sqrt{ax + x^2}},$$

so that

$$\int \frac{x\,dx}{\sqrt{ax + x^2}} = \sqrt{ax + x^2} - \frac{1}{2} a \log. \left(x + \frac{1}{2} a + \sqrt{ax + x^2}\right) + C.$$

3. Required the integral of

$$\sqrt{\frac{a - x}{x}}\, dx.$$

Multiplying numerator and denominator by $\sqrt{a - x}$ this becomes

$$\frac{a\,dx - x\,dx}{\sqrt{ax - x^2}},$$

which would be the differential of $\sqrt{ax - x^2}$ if the numerator were diminished by $\frac{1}{2} a\,dx$; hence, first subtracting and then adding this quantity, we have

$$\int \sqrt{\frac{a - x}{x}}\, dx = \int \frac{\frac{1}{2} a\,dx - x\,dx}{\sqrt{ax - x^2}} + \int \frac{\frac{1}{2} a\,dx}{\sqrt{ax - x^2}}$$

$$= \sqrt{ax - x^2} + \frac{1}{2} a \text{ versin.}^{-1} \frac{2x}{a} + C.$$

4. Required the integral of

$$\frac{(1 - x^2)\, dx}{(1 + x^2) \sqrt{1 + ax^2 + x^4}}.$$

Divide both numerator and denominator by x^2 and the expressio becomes

$$\frac{\dfrac{1 - x^2}{x^2}\, dx}{\dfrac{1 + x^2}{x} \sqrt{\dfrac{1}{x^2} + a + x^2}}.$$

Put

$$y = \frac{1 + x^2}{x} \quad \therefore dy = - \frac{1 - x^2}{x^2}\, dx,$$

also

$$y^2 - 2 = \frac{1}{x^2} + x^2,$$

consequently the proposed differential becomes transformed into

$$- \frac{dy}{y \sqrt{y^2 + a - 2}},$$

of which the integral by ex. 8, p. 34, is

$$\frac{1}{\sqrt{a - 2}}\, \log. \frac{\sqrt{y^2 + a - 2} + \sqrt{a - 2}}{y}.$$

5. Required the integral of

$$\frac{1 + x^2}{(1 - x^2)\sqrt{1 + x^4}}\, dx.$$

Assume

$$y(1 - x^2) = 1 + x^2 \quad \therefore x^2 = \frac{y - 1}{y + 1}$$

$$\therefore dx = \frac{dy}{(y - 1)^{\frac{1}{2}}(y + 1)^{\frac{3}{2}}}.$$

Also

$$1 + x^4 = 1 + \frac{(y - 1)^2}{(y + 1)^2} = 2\,\frac{y^2 + 1}{(y + 1)^2};$$

hence, by substituting these values in the proposed expression, t
becomes

$$\frac{y\, dy}{2^{\frac{1}{2}}(y^4 - 1)^{\frac{1}{2}}} = \frac{1}{2^{\frac{3}{2}}} \cdot \frac{2\, y\, dy}{(y^4 - 1)^{\frac{1}{2}}} = \frac{dz}{\sqrt{z^2 - 1}},$$

z being put for y^2.   The integral of this is (18)

$$\frac{1}{2^{\frac{3}{2}}}\, \log. \{\sqrt{z^2 - 1} + z\} + C =$$

$$\frac{1}{2^{\frac{3}{2}}}\, \log. \{\sqrt{y^4 - 1} + y^2\} + C =$$

$$\frac{1}{2^{\frac{3}{2}}}\log.\ \{[\frac{(1+x^2)^4}{(1-x^2)^4}-1]^{\frac{1}{2}}+\frac{(1+x^2)^2}{(1-x^2)^2}\}+C.$$

6. Required the integral of

$$\sqrt{1+\cos.\ x}\ .\ dx.$$

Put

$$1+\cos.\ x=z\ \therefore-\sin.\ x\ dx=dz,$$

therefore

$$-\frac{\sin.\ x\ dx}{\sqrt{1-\cos.\ x}}=\frac{dz}{\sqrt{2-z}},$$

but

$$\frac{\sin.\ x}{\sqrt{1-\cos.\ x}}=\frac{\sin.\ x\sqrt{1+\cos.\ x}}{\sqrt{1-\cos.^2 x}}=\frac{\sin.\ x\sqrt{1+\cos.\ x}}{\sin.\ x}$$

$$=\sqrt{1+\cos.\ x}\ ;$$

hence

$$\int\sqrt{1+\cos.\ x}\ dx=-\int\frac{dz}{\sqrt{2-z}}=2\sqrt{2-z}+C,$$

that is

$$\int\sqrt{1+\cos.\ x}\ dx=2\sqrt{1-\cos.\ x}+C.$$

7. Required the integral of

$$\sin.\ mx\ \sin.\ nx\ dx.$$

Since by Trigonometry

$$\cos.\ (A-B)=\cos.\ A\ \cos.\ B+\sin.\ A\ \sin.\ B$$

$$\cos.\ (A+B)=\cos.\ A\ \cos.\ B-\sin.\ A\ \sin.\ B$$

$$\therefore\frac{1}{2}\cos.\ (A-B)-\frac{1}{2}\cos.\ (A+B)=\sin.\ A\ \sin.\ B,$$

or, substituting mx for A and nx for B, we have

$$\sin.\ mx\ \sin.\ nx=\frac{1}{2}\cos.\ (m-n)x\ \ \frac{1}{2}\cos.\ (m+n)x\ ;$$

hence, multiplying by dx and integrating, we get

$$\int\sin.\ mx\ \sin.\ nx\ dx=$$

$$\frac{1}{2} \int \cos. (m-n) x \, dx - \frac{1}{2} \int \cos. (m+n) x \, dx$$

$$= \frac{1}{2} \left\{ \frac{\sin. (m-n) x}{m-n} - \frac{\sin. (m+n) x}{m+n} \right\} + C.$$

8. Required the integral of

$$\cos. mx \cos. nx \, dx.$$

By adding together the two trigonometrical formulas above, we have

$$\frac{1}{2} \cos. (A-B) + \frac{1}{2} \cos. (A+B) = \cos. A . \cos. B;$$

hence

$$\cos. mx \cos. nx = \frac{1}{2} \cos. (m-n) x + \frac{1}{2} \cos. (m+n) x,$$

therefore multiplying by dx and integrating

$$\int \cos. mx \cos. nx \, dx =$$

$$\frac{1}{2} \left\{ \frac{\sin. (m-n) x}{m-n} + \frac{\sin. (m+n) x}{m+n} \right\}.$$

In like manner, by adding together the expressions for sin. $(A + B)$ and sin. $(A - B)$, we get

$$\int \sin. mx \cos. nx \, dx =$$

$$- \frac{1}{2} \left\{ \frac{\cos. (m+n) x}{m+n} + \frac{\cos. (m-n) x}{m-n} \right\}.$$

9. Required the integral of

$$\frac{dx}{a + b \sin. x}.$$

Substitute y for sin. x, then

$$dx = \frac{dy}{\sqrt{1-y^2}},$$

therefore the expression becomes

$$\frac{dy}{(a+by)\sqrt{1-y^2}},$$

which may be rendered rational by (17), or, putting

$$1 - y^2 = (1 - y)^2 z^2 \therefore y = \frac{z^2 - 1}{z^2 + 1} \therefore dy = \frac{4z\, dz}{(z^2 + 1)^2}$$

the expression becomes

$$\frac{2\, dz}{a - b + (a + b)\, z^2},$$

which is rational.

10. To determine the integral of

$$\frac{A x^2\, dx}{\sqrt{P x^2 - x^4 - Q}}.$$

This may be done by developing the denominator in a series, but there is a neater and much more commodious manner of expressing the integral, since the proposed differential may be assimilated to that of an elliptic arc. Thus, calling s the length of an arc of an ellipse, we have (61)

$$ds = \frac{\sqrt{a^2 - \varepsilon^2 x^2}}{\sqrt{a^2 - x^2}}\, dx,$$

and if in this expression we put

$$a^2 - \varepsilon^2 x^2 = x'^2 \text{ or } a^2 - \frac{a^2 - b^2}{a^2} x^2 = x'^2$$

we shall have

$$x = \frac{a \sqrt{a^2 - x'^2}}{\sqrt{a^2 - b^2}}$$

$$\therefore dx = \frac{a x'\, dx'}{\sqrt{a^2 - b^2} \sqrt{a^2 - x'^2}}, \quad \sqrt{a^2 - x^2} = \frac{a \sqrt{x'^2 - b^2}}{\sqrt{a^2 - b^2}}$$

$$\therefore ds = \frac{x'^2\, dx'}{\sqrt{(a^2 + b^2)\, x'^2 - x'^4 - a^2 b^2}}.$$

Hence the integral sought may be expressed by the product of a constant factor A by the arc of an ellipse, of which the abscissa of the extremity is x. The axes of the ellipse are found by equating the expression under the radical here with that in the proposed differential, which leads to

$$a^2 + b^2 = P, \ a^2 b^2 = Q,$$

whence the axes are

$$2a = \sqrt{P + 2\sqrt{Q}} + \sqrt{P - 2\sqrt{Q}}$$
$$2b = \sqrt{P + 2\sqrt{Q}} - \sqrt{P - 2\sqrt{Q}},$$

so that the integral of the proposed differential is A times an elliptic arc of which these are the axes, and of which the abscissa of the extremity is

$$\frac{a\sqrt{a^2 - x^2}}{\sqrt{a^2 - b^2}}.$$

It must however be observed that this is not the case unless $P > 2\sqrt{Q}$, for if $P < 2\sqrt{Q}$ the axes become imaginary, and if $P = 2\sqrt{Q}$, then

$$\sqrt{Px^2 - x^4 - Q} \quad \text{or} \quad \sqrt{\left(\tfrac{1}{4}P^2 - Q\right) - \left(x^2 - \tfrac{1}{2}P\right)^2}$$

is obviously imaginary.

11. To determine the integral of

$$\frac{A\,dx}{x^2\sqrt{Px^2 - x^4 - Q}}.$$

In the expression

$$ds = \frac{\sqrt{a^2 - \varepsilon^2 x^2}}{\sqrt{a^2 - x^2}}\,dx$$

put

$$a^2 - \varepsilon^2 x^2 = \frac{a^2 b^2}{x'^2} \quad \text{or} \quad a^2 - \frac{a^2 - b^2}{a^2} = \frac{a^2 b^2}{x'^2}$$

and we get

$$x = \frac{a^2\sqrt{x'^2 - b^2}}{x'\sqrt{a^2 - b^2}},$$

whence

$$ds = \frac{-a^2 b^2\,dx'}{x'^2\sqrt{(a^2 + b^2)x'^2 - x'^4 - a^2 b^2}};$$

therefore, comparing this with the proposed, which we may write

$$-\frac{A}{Q} \times \frac{-Q\,dx}{x^2\sqrt{Px^2 - x^4 - Q}},$$

we see that the sought integral is $-\dfrac{A}{Q}$ times an elliptic arc, the abscissa of whose extremity is

$$\frac{a^2 \sqrt{x^2 - b^2}}{x \sqrt{a^2 - b^2}},$$

and whose axes, as determined from the conditions

$$a^2 + b^2 = P, \; a^2 b^2 = Q,$$

are

$$2a = \sqrt{P + 2\sqrt{Q}} + \sqrt{P - 2\sqrt{Q}}$$

$$2b = \sqrt{P + 2\sqrt{Q}} - \sqrt{P - 2\sqrt{Q}}.$$

12. To determine the integral of

$$\frac{A x^2 \, dx}{\sqrt{P x^2 + x^4 - Q}}.$$

The equation of the hyperbola is

$$y^2 = (\varepsilon^2 - 1)(x^2 - a^2,$$

in which

$$\varepsilon^2 = \frac{a^2 + b^2}{a^2}.$$

therefore, if s represent any arc measured from the vertex of the transverse axis, we have

$$ds = \frac{\sqrt{\varepsilon^2 x^2 - a^2}}{\sqrt{x^2 - a^2}} \, dx.$$

Assume

$$\varepsilon^2 x^2 - a^2 = x'^2 \; \therefore \; x = \frac{\sqrt{x'^2 + a^2}}{\varepsilon};$$

hence

$$ds = \frac{x' \, dx'}{\varepsilon \sqrt{x'^2 + a^2}} \cdot \frac{x'}{\sqrt{\dfrac{x'^2 + a^2}{\varepsilon} - a^2}}$$

$$= \frac{x'^2 \, dx'}{\sqrt{(a^2 - b^2) x'^2 + x'^4 - a^2 b^2}},$$

which agrees with the proposed form. Hence the integral sought is A times the arc of an hyperbola whose abscissa is

$$\frac{\sqrt{x^2 + a^2}}{\varepsilon},$$

and whose axes, determined from the conditions

$$a^2 - b^2 = P, \; a^2 b^2 = Q,$$

are

$$2a = \sqrt{2P + 2\sqrt{P^2 + 4Q}}$$

$$2b = \sqrt{-2P + 2\sqrt{P^2 + 4Q}}.$$

13. To determine the integral of

$$\frac{A\,dx}{x^2 \sqrt{P x^2 - x^4 + Q}}.$$

In the expression for ds, in last example, put

$$\varepsilon^2 x^2 - a^2 = \frac{a^2 b^2}{x'^2} \; \therefore \; x = \frac{a\sqrt{x'^2 + b^2}}{\varepsilon x'}$$

and it will become

$$ds = \frac{-a^2 b^2\,dx'}{x'^2 \sqrt{(a^2 - b^2) x'^2 - x'^4 + a^2 b^2}},$$

comparing this with the proposed expression when written

$$-\frac{A}{Q} \times \frac{-Q\,dx}{x^2 \sqrt{P x^2 - x^4 + Q}}$$

we find for the sought integral $-\dfrac{A}{Q}$ times an arc of the hyperbola whose axes are

$$2a = \sqrt{2P + 2\sqrt{P^2 + 4Q}}$$

$$2b = \sqrt{-2P + 2\sqrt{P^2 + 4Q}},$$

and abscissa

$$\frac{a\sqrt{x^2+b^2}}{\varepsilon\,x}=\frac{a^2\sqrt{x^2+b^2}}{x\sqrt{a^2+b^2}}.$$

14. To determine the integral of

$$\sqrt{A+\frac{B}{C-x^2}}\cdot dx.$$

Every differential of this kind may be compared to that of an elliptic arc, since it may be reduced to the form

$$\frac{\sqrt{a^2-\varepsilon^2 x^2}}{\sqrt{a^2-x^2}}\,dx.$$

For the proposed form is the same as

$$\frac{\sqrt{AC+B-Ax^2}}{\sqrt{C-x^2}}\,dx=\sqrt{A+\frac{B}{C}}\left\{\frac{\sqrt{\left(C-\dfrac{CA}{AC+B}x^2\right)}}{\sqrt{C-x^2}}\right\}dx,$$

which, integrated, gives $\sqrt{A+\dfrac{B}{C}}$ times an elliptic arc, whose abscissa is x, major semi axis $\sqrt{C}$, and excentricity

$$C\sqrt{\frac{A}{AC+B}}.$$

In a similar manner may

$$\sqrt{A+\frac{B}{x^2-C}}\cdot dx$$

be integrated by means of an hyperbolic arc.

15. To determine the integral of

$$\frac{\sqrt{A+By^2}}{\sqrt{C\pm y^2}}\,dy,$$

If we determine the expression for ds according to the second of the general formulas in (59) we shall find that when s is the arc of an ellipse or of an hyperbola

$$ds=\frac{\sqrt{b^4+a^2 e^2 y^2}}{\sqrt{b^2\mp y^2}}\cdot\frac{dy}{b},$$

the upper sign for the ellipse, the lower for the hyperbola, and to this form the proposed may be assimilated by writing it thus:

$$\frac{\sqrt{A}}{C}\left\{\frac{\sqrt{C^2 + \frac{C^2 B}{A} y^2}}{\sqrt{C \pm y^2}}\right\} dy,$$

consequently the integral sought is $\dfrac{\sqrt{A}}{C}$ times an elliptic or hyperbolic arc, of which the ordinate of the extremity is y, and of which the conjugate semi axis is $\sqrt{C}$, and excentricity, $c = C \dfrac{\sqrt{B}}{\sqrt{A}}$, that is, the two semi axes are

$$\sqrt{C} \text{ and } \sqrt{C^2 B \pm AC},$$

the upper sign having place if the arc is elliptic, and the lower if it is hyperbolic.

SECTION II.

ON THE

INTEGRATION OF DIFFERENTIAL EXPRESSIONS OF SEVERAL VARIABLES.

CHAPTER I.

INTEGRATION OF EXACT DIFFERENTIALS.

(77.) Any expression involving variable quantities and their differentials is called an *exact differential*, when it is immediately derivable from some function of those variables by the common process of differentiation. In such cases the primitive function is always readily determinable, or rather the integration may be always made to depend on that of a differential expression of a single variable. It becomes of consequence, therefore, when any differential expression is proposed, to be able to ascertain, first, whether it be an exact differential, and second, if it be exact, how to discover the primitive function. It will be the object of the present chapter to shew how these objects are to be accomplished.

Euler's Criterion of Integrability.

(78.) Let the proposed differential expression be

$$M dx + N dy,$$

in which M and N are functions of x and y. If this is an exact differential of any primitive function u, it must have arisen from differentiating u relatively to both the variables x, y. But

$$du = \frac{du}{dx} dx + \frac{du}{dy} dy;$$

hence, if our supposition is correct, we must have the equations

$$M = \frac{du}{dx}, \; N = \frac{du}{dy}.$$

These, then, are the conditions from which we are to determine whether or not the proposed is really an exact differential, and they are quite sufficient for this purpose, inasmuch as they immediately lead to the necessary relation between M and N. For, let us differentiate these two equations, the first relatively to y, the second relatively to x, and we shall have

$$\frac{dM}{dy} = \frac{d^2u}{dx\,dy}, \quad \frac{dN}{dx} = \frac{d^2u}{dy\,dx},$$

but, (*Diff. Calc.* p. 85,)

$$\frac{d^2u}{dx\,dy} = \frac{d^2u}{dy\,dx};$$

hence, that the proposed may be an exact differential, there must exist this relation between the functions M and N, viz.

$$\frac{dM}{dy} = \frac{dN}{dx},$$

which is therefore called *the criterion of integrability.*

If the proposed differential contained three variables as

$$Mdx + Ndy + Pdz,$$

then, as before, assuming the primitive function to be u, we have, since

$$du = \frac{du}{dx} dx + \frac{du}{dy} dy + \frac{du}{dz} dz,$$

the conditions

$$M = \frac{du}{dx}, \; N = \frac{du}{dy}, \; P = \frac{du}{dz}$$

consequently

$$\frac{dM}{dy} = \frac{dN}{dx}$$

$$\frac{dM}{dz} = \frac{dP}{dx}$$

$$\frac{d\mathrm{N}}{dz} = \frac{d\mathrm{P}}{dy}$$

are the necessary conditions of integrability, and it is obvious that generally for every differential function containing n variables, we must have $\dfrac{n(n-1)}{2}$ such equations of condition, if the proposed is an exact differential, because this formula expresses the number of combinations of every two of the n variables.

Suppose the given differential were

$$(3x^2 + 2axy)\,dx + (ax^2 + 3y^2)\,dy,$$

then, since

$$\frac{d\mathrm{M}}{dy} = 2ax, \quad \frac{d\mathrm{N}}{dx} = 2ax,$$

we may be sure that the differential proposed is exact.

Again, let

$$\frac{(3x^2 - y)\,dx - x\,dy}{2\sqrt{x^3 - xy}}$$

be proposed, then

$$\frac{d\mathrm{M}}{dy} = d\,\frac{3x^2 - y^2}{2\sqrt{x^3 - xy}} \div dy = \frac{x^3 + xy}{2\sqrt{x^3 - xy}}$$

$$\frac{d\mathrm{N}}{dx} = d\,\frac{-x}{2\sqrt{x^3 - xy}} \div dx = \frac{x^3 + xy}{2\sqrt{x^3 - xy}}$$

so that this also is an exact differential.

But, if

$$(x^2 y + y^2)\,dx - x^3\,dy + xy^2\,dy$$

be the proposed differential, then, since

$$\frac{d\mathrm{M}}{dy} = x^2 + 2y, \quad \frac{d\mathrm{N}}{dx} = y^2 - 3x^2,$$

we should infer that the differential in question cannot arise from *immĕdiately* differentiating any function whatever.

Let us now proceed to the integration of differential expressions, which satisfy the foregoing conditions.

Integration of Differentials which satisfy the Conditions of Integrability.

(79.) As the first term Mdx of the exact differential

$$du = Mdx + Ndy$$

has been obtained by differentiating the primitive function u, as if y were a constant, it follows that if we integrate this first term on the same hypothesis, the integral properly corrected must be the original function u, that is

$$u = \int Mdx + C,$$

observing, however, that in consequence of y being considered constant, it may enter the correction C, so that, as C may be a function of y, it will be better to write this expression thus:

$$u = \int Mdx + Y \ldots (1).$$

By applying similar reasoning to Ndy, the second term of the proposed differential, we should obtain for u the expression

$$u = \int Ndy + X \ldots (2),$$

where X the correction may contain x. It merely remains, therefore, to determine the proper correction Y or X in one of the equations (1), (2). Let us take the first then, since $N = \dfrac{du}{dy}$, it follows from (1) that

$$N = \frac{d \int Mdx}{dy} + \frac{dY}{dy}$$

$$\therefore \frac{dY}{dy} = N - \frac{d \int Mdx}{dy}$$

$$\therefore Y = \int \left(N - \frac{d \int Mdx}{dy} \right) dy.$$

The quantity within the parentheses cannot possibly contain x, otherwise this conclusion could not be deduced; it would, indeed, be contradictory, seeing that Y cannot contain x*. Substituting in (1) this expression for Y, we have, for the complete integral sought,

$$u = \int M dx + \int \left(N - \frac{d \int M dx}{dy}\right) dy \ldots (3).$$

which is, therefore, a general formula for the integration of exact differentials of two variables.

If we had taken equation (2) instead of equation (1) a similar process would have led to the general formula

$$u = \int N dy + \int \left(M - \frac{d \int N dy}{dx}\right) dx \ldots (4.)$$

We shall now add an example or two of the application of these general formulas.

EXAMPLES.

1. To determine the integral of

$$(6xy - y^2)\, dx + (3x^2 - 2xy)\, dy.$$

In this example,

$$M = 6xy - y^2, \ N = 3x^2 - 2xy$$

$$\therefore \frac{dM}{dy} = 6x - 2y = \frac{dN}{dx},$$

the differential proposed is, therefore, exact, and consequently the integral is comprised in the general formula (3) or (4). Instead, however, of availing ourselves of this formula, we shall employ the process

---

* This condition is, in fact, involved in, and depends upon, that of integrability, for, by differentiating the expression within the parenthesis relatively to x, the result is

$$\frac{dN}{dx}\, dx - \frac{dM}{dy}\, dx,$$

which, by the condition of integrability, is 0.

which led to it, in order to render that process more familiar to the student.

Integrating the terms containing dx on the hypothesis that y is constant, we have

$$\int (6xy - y^2)\, dx = 3x^2 y - y^2 x + Y,$$

and this, when Y is determined, must be the integral sought, and it is the differential of this integral with respect to y that forms the term containing dy in the proposed; hence, then, differentiating the expression just deduced, with respect to y, we must have the identity

$$3x^2 - 2xy + \frac{dY}{dy} = 3x^2 - 2xy$$

$$\therefore \frac{dY}{dy} = 0 \therefore Y = \text{constant};$$

hence the integral of the proposed expression is

$$3x^2 y - y^2 x + C.$$

2. To determine the integral of

$$(2y^2 x + 9x^2 y + 8x^3)\, dx + (2x^2 y + 3x^3)\, dy.$$

Here

$$\frac{dM}{dy} = 4yx + 9x^2 = \frac{dN}{dx},$$

so that the differential is exact.

Integrating with respect to x, we have

$$\int (2y^2 x + 9x^2 y + 8x^3)\, dx = y^2 x^2 + 3x^3 y + 2x^4 + Y,$$

and differentiating this with respect to y, and then comparing it to the term containing dy in the proposed, we have

$$2yx^2 + 3x^3 + \frac{dY}{dy} = 2x^2 y + 3x^3$$

$$\therefore \frac{dY}{dy} = 0 \therefore Y = C;$$

hence the integral sought is

$$y^2 x^2 + 3x^3 y + 2x^4 + C.$$

3. To determine the integral of

$$\frac{dx}{\sqrt{1 + x^2}} + adx + 2by\,dy.$$

This is obviously an exact differential, since y does not enter into the coefficient of dx, nor x into the coefficient of dy. Hence, integrating the term containing dy, which is the simplest, we have,

$$2\int by\,dy = by^2 + \mathrm{X},$$

and equating the differential of this, with respect to x, with the other terms of the proposed, we get

$$d\mathrm{X} = \frac{dx}{\sqrt{1 + x^2}} + adx$$

$$\therefore \mathrm{X} = \log.\,(x + \sqrt{1 + x^2}) + ax + \mathrm{C};$$

hence the required integral is

$$by^2 + \log.\,(x + \sqrt{1 + x^2}) + ax + \mathrm{C}.$$

We shall give one example of the process as applied to a differential expression of three variables.

4. To determine the integral of

$$du = \frac{ydz}{z} + \frac{(x + 2ay)\,dy}{z} - \frac{(xy + ay^2)\,dz}{z^2}.$$

Here

$$\mathrm{M} = \frac{y}{z},\ \mathrm{N} = \frac{x + 2ay}{z},\ \mathrm{P} = -\frac{xy + ay^2}{z^2}$$

and

$$\frac{d\mathrm{M}}{dy} = \frac{1}{z} = \frac{d\mathrm{N}}{dx}\ ,\ \frac{d\mathrm{M}}{dz} = -\frac{y}{z^2} = \frac{d\mathrm{P}}{dx},$$

$$\frac{d\mathrm{N}}{dz} = -\frac{x + 2ay}{z^2} = \frac{d\mathrm{P}}{dy};$$

hence the proposed is an exact differential; so that $\dfrac{ydx}{z}$ is the partial differential of u, taken relatively to the single variable x; hence, integrating with respect to this variable, we have

$$u = \frac{y}{z} \int dx = \frac{yx}{z} + \mathrm{F}\,(y.z) \;\ldots\ldots\; (1);$$

it remains therefore, to determine the function $\mathrm{F}\,(y, z)$, which completes this integral. Differentiating with respect to y, and equating the resulting coefficient with that of dy, in the proposed we have

$$\frac{du}{dy} = \frac{x}{z} + \frac{d\,\mathrm{F}\,(x, y)}{dy} = \frac{x + 2ay}{z}$$

$$\therefore \frac{d\,\mathrm{F}\,(x, y)}{dy} = \frac{2ay}{z}$$

$$\therefore \mathrm{F}\,(x, y) = \frac{2a}{z}\int y\,dy = \frac{ay^2}{z} + fz.$$

Substituting this expression for $\mathrm{F}\,(x, y)$, in the equation (1), it becomes

$$u = \frac{yx}{z} + \frac{ay^2}{z} + fz \;\ldots\ldots\; (2).$$

We have, therefore, now only to determine the correction fz, which is effected by differentiating with respect to z, and equating the coefficient with that of dz in the proposed, so that we have

$$\frac{du}{dz} = -\frac{yx}{z^2} - \frac{ay^2}{z^2} + \frac{dfz}{dz} = -\frac{xy + 2ay}{z^2}$$

$$\therefore \frac{dfz}{dz} = 0 \;\therefore fz = \mathrm{C};$$

hence equation (2) becomes

$$u = \frac{yx + ay^2}{z} + \mathrm{C},$$

the integral sought.

5. To determine the integral of

$$du = x\,dy + y\,dx - \frac{dy}{xy^2} - \frac{dx}{yx^2}$$

$$u = yx + \frac{1}{yx}.$$

6. To determine the integral of

$$du = (ax + by + c)\,dx + (bx + my + n)\,dy$$

$$u = \frac{1}{2} ax^2 + byx + cx + \frac{1}{2} my^2 + ny + C.$$

7. To determine the integral of

$$du = \frac{dx}{\sqrt{x^2 + y^2}} + \left\{1 - \frac{x}{\sqrt{x^2 + y^2}}\right\} \frac{dy}{y}$$

$$u = x + \sqrt{x^2 + y^2} + C.$$

Differentials which are both Exact and Homogeneous.

(80.) An algebraical function, consisting of several terms, is said to be *homogeneous*, when the sum of the exponents of the variables is the same in every term.

Thus the following are homogeneous functions, viz.

$$ax^3 y^2 z + by x^5, \quad \frac{ax^3 y^3 + y^6}{\sqrt{x^4 - y^4}}, \quad \frac{axy + y^2 + z^2}{\sqrt{x^6 - y^6}},$$

the *degree* of homogeneity being in the first 6, in the second $6 - 2 = 4$, and in the third $2 - 3 = - 1$.

When exact differentials are also homogeneous, their integrals may be obtained by a very easy process, except in that particular case where the degree of homogeneity is $- 1$, in which case the process we are going to explain is not always applicable.

Let

$$A dx + B dy + C dz + \&c. \ \ldots \ldots (1)$$

be an exact differential, and such that A, B, C, &c. are homogeneous functions of the variables, the degree of homogeneity being n, then the primitive, u, of this differential will be

$$u = \frac{Ax + By + Cz + \&c.}{n + 1}.$$

For suppose, in the primitive function u, that instead of y, z, &c. there be substituted $y'x$, $z'x$, &c. then, in consequence of the degree of homogeneity of this function being $n + 1$, it will become divisible by x^{n+1}, so that we may represent it by

$$u = Px^{n+1} \ \ldots \ (2),$$

P being a function of y', z', &c. Also the proposed differential will be divisible by x^n, and may, therefore, be represented by

$$A'x^n \, dx + B'x^n \, d \, . \, y'x + C' \, x^n \, d \, . \, z'x + \&c. \ \ldots \ (3).$$

Let us now differentiate (2) relatively to x, only then

$$\frac{du}{dx} \, dx = (n + 1) \, Px^n \, dx \ \ldots \ (4),$$

and since (3) is the total differential of u, we shall obtain the partial differential relatively to x, by suppressing all the terms connected with dy', dz', &c.; that is,

$$\frac{du}{dx} \, dx = A'x^n \, dx + B'x^n \, y' \, dx + C \, x^n \, z' dx + \&c. \ \ldots \ (5).$$

Divide each of the identities (4), (5), by $\dfrac{dx}{x}$, and we have

$$(n + 1) \, Px^{n+1} = A'x^{n+1} + B'x^{n+1} \, y' + C'x^{n+1} \, z' + \&c.$$

Restoring now the values of y', z', &c. viz.

$$y' = \frac{y}{x}, \ z' = \frac{z}{x}, \ \&c.$$

and recollecting that

$$A'x^n = A, \ B'x^n = B, \ \&c.$$

this equation is the same as

$$(n + 1) \, Px^{n+1} = Ax + By + Cz + \&c.$$

therefore ·

$$Px^{n+1} = u = \frac{Ax + By + Cz + \&c.}{n + 1}$$

It hence appears that to determine the integral of (1) it is necessary merely to change dx, dy, dz, &c. and to divide by the index of homogeneity increased by unity. When the index n is -1, then the divisor is $1 - 1$, and this process is liable to exception, for to such differentials belong those of logarithms and circular arcs, and, although these differentials are themselves free from transcendental quantities, and have, therefore, a determinate degree of homogeneity n, yet it is not true of such that their primitives have the degree $n + 1$, so that for such differentials the foregoing process is inapplicable.

(81.) We shall now give an example or two of this method:

EXAMPLES.

1. To integrate

$$du = (3x^2 + 2axy)\, dx + (ax^2 + 3y^2)\, dy.$$

This expression fulfils the condition of integrability; hence, by the foregoing rule,

$$u = \frac{3x^3 + 2ax^2y + ax^2y + 3y^3}{2 + 1} = x^3 + ax^2y + y^3 + C,$$

which is the integral required.

2. To integrate

$$(3x^2 + 2bxy - 3y^2)\, dx + (bx^2 - 6xy + 3cy^2)\, dy.$$

This differential also is both exact and homogeneous; hence by the rule the integral is

$$\frac{3x^3 + 2bx^2y - 3xy^2 + bx^2y - 6xy^2 + 3cy^3}{3} =$$

$$x^3 + bx^2y - 3xy^2 + cy^3 + C.$$

When homogeneous differentials are proposed for integration, it is often easier to apply the foregoing method of integration at once, without first trying the criterion of integrability, and then, by differentiating the result, we shall return to the proposed, if this be an exact differential, otherwise the expression is not immediately integrable. If the homogeneous differential consist of three or four different variables, this mode of proceeding will be decidedly preferable. To exemplify this *M. Dubourguet* adduces the following:

3. To integrate

$$\frac{2 \sqrt{xy - z}}{2u^2 \sqrt{x}}\, dx + \frac{xdy}{2u^2 \sqrt{y}} -$$

$$2 \frac{x \sqrt{y} - z \sqrt{x}}{u^3} du - \frac{\sqrt{x} \, dz}{u^2},$$

of which the degree of homogeneity is $-\dfrac{3}{2}$.

By substituting, according to the rule x, y, z, and u, instead of their differentials dx, dy, dz, and du, and dividing the result by $-\dfrac{3}{2} + 1 = -\dfrac{1}{2}$, we find

$$-\frac{2x \sqrt{y}}{u^2} + \frac{z \sqrt{x}}{u^2} - \frac{x \sqrt{y}}{u^2} + \frac{4x \sqrt{y}}{u^3} - \frac{4z \sqrt{x}}{u^2} + \frac{2z \sqrt{x}}{u^2} =$$

$$\frac{x \sqrt{y} - z \sqrt{x}}{u^2} + C;$$

this result, differentiated, produces the proposed differential, which is therefore thus integrated. But if the differential of the result had not agreed with the proposed, the trouble of thus ascertaining this would be much less than that of seeking the six equations of condition in art. (78).

4. To determine the integral of

$$du = (2y^2 x + 3y^3) \, dx + (2x^2 y + 9xy^2 + 8y^3) \, dy$$
$$u = y^2 x + 3y^3 x + 2y^4 + C.$$

5. To determine the integral of

$$du = \frac{y dx}{z} + \frac{(x - 2y) \, dy}{z} + \frac{(y^2 - xy) \, dz}{z^2}$$
$$u = \frac{xy - y^2}{z} + C.$$

The variables which enter the several differential expressions integrated in the present chapter are considered to be entirely independent, and the expressions themselves to be independent of other expressions, so that, if they do not satisfy the conditions of integrability, we must consider them as altogether unintegrable, since we are not at liberty to perform any preliminary operation upon them that might render them integrable. If, however, we have a relation between any two differential expressions, so that we may obtain an equation between them, then, as the equation still subsists, whatever operations we perform on each of its members, we may obviously use means to render the equation integrable, without altering the relation which the equation fixes among the variables.   It is thus that the integration of differential equations is a much more extensive, as well as important, subject of inquiry, than the integration of isolated expressions, and it is this subject that will occupy us during the remaining part of the present volume.

———

CHAPTER II.

ON THE THEORY OF DIFFERENTIAL EQUATIONS AND OF ARBITRARY CONSTANTS.

(82.) Let

$$u = F(x, y) = 0 \ldots \ldots (1)$$

be an equation, cleared of radicals, between two variables, whose relation to each other is thus fixed. If we differentiate this equation successively, x being the independent variable, we shall have the series of equations

$$\frac{du}{dx} = 0, \ \frac{d^2u}{dx^2} = 0, \ \frac{d^3u}{dx^3} = 0, \ \&c. \ldots \ldots (2),$$

in all of which the same relation between x and y subsists as in the primitive equation, for differentiation does not alter this relation, so that these equations all exist simultaneously with the primitive.

Now we may remark of these differential equations that the first, $\dfrac{du}{dx} = 0$,

is a function of x, y, and $\dfrac{dy}{dx}$, this last entering only in the first power, and that the equation contains the very same constants as the primitive, with the exception of that which has disappeared by the process of differentiation.. We shall consider such a constant as this latter to have actually entered the function (1), as we propose to view not only $\frac{du}{dx} = 0$ as a differential function derived from (1), but moreover that

(1) is the *complete* primitive of $\frac{du}{dx} = 0$. In like manner, the equation

$\frac{d^2u}{dx^2} = 0$, which is a function of $x, y, \frac{dy}{dx}$, and $\frac{d^2y}{dx^2}$, the latter entering only in the first power, contains the same constants as its primitive $\frac{du}{dx} = 0$, with the exception of that which has disappeared by differen-

tiation, and this constant we shall suppose to enter $\frac{du}{dx} = 0$, viewing it

as the complete primitive of $\frac{d^2u}{dx^2} = 0$. If, then, we stop at the diffe

rential equation of the nth order, viz. $\frac{d^nu}{dx^n} = 0$, considering all along each to be the complete primitive of that which immediately succeeds, it follows that the original function $F(x, y) = 0$ must contain n arbi-

trary constants, and will be the complete final integral of $\frac{d^nu}{dx^n} = 0$.

Now, as there are $n - 1$ constants common to the two equations $F(x, y) = 0$ and $\frac{du}{dx} = 0$, we may eliminate any one of these, and may thus

obtain $n - 1$ new differential equations of the first order, or containing

no higher differential coefficient than $\frac{dy}{dx}$. These equations are neces-

sarily all different, for at every elimination the entire term connected with the constant has been eliminated. The same relation, however, subsists between x and y in each of these equations, viz. the relation (1); this, therefore, is equally the complete primitive of either of them, and they can have no other. It will obviously be obtained by eliminating $\dfrac{dy}{dx}$ from any two of the n differential equations of the first order, since the result will be the relation between x and y, in virtue of which they simultaneously exist. Each of the $n - 1$ equations which have been obtained by elimination may also, by modifying the primitive, be obtained by differentiation as the immediate differential was in the first instance. For, if the original primitive be solved for any one of the constants A, and we find $A = f(x, y)$ then the equation $f(x, y) - A = 0$ must be the same primitive under a different form, and if this be differentiated, the constant A will disappear, while all the others will enter the resulting differential equation.

This differential equation, however, is not precisely the same as that arising from eliminating the constant A, although immediately reducible to it by the introduction of a factor, and this it is of consequence to prove.

In order to put in evidence the constant A, which we wish to climinate, let us write the primitive in this manner,

$$F(x, y) + Af(x, y) = 0 \ldots \ldots (1),$$

then, for the immediate differential, we shall have

$$dF(x, y) + Adf(x, y) = 0,$$

and if we eliminate Λ by means of these two equations, the result will be the differential equation

$$F(x, y)\, df(x, y) - f(x, y)\, dF(x, y) = 0 \ldots \ldots (2).$$

Now this equation is not precisely the same as that which would arise from solving (1) for Λ, and differentiating the result, although it may be readily rendered so by a factor. For the equation (1), when solved for A, is

$$A = -\frac{F(x, y)}{f(x, y)} \ldots \ldots (3),$$

of which the immediate differential is

$$\frac{F(x, y)\, df(x, y) - f(x, y)\, dF(x, y)}{\{f(x, y)\}^2} = 0 \ldots \ldots (4).$$

Hence, as affirmed above, (2) is not the immediate or *exact differential* of (3), although it may be rendered so, by introducing the factor

$$\frac{1}{\{f(x, y)\}^2} \cdot \cdot \cdot \cdot (5).$$

The equations (2) and (4) both involve the same relation (1) or (3) between the variables, yet we could not return from (2) to (1) or (3) without first introducing the factor (5), as this is necessary, in order to render it an exact differential.

It must be remarked here that if besides A there also enter powers of it in the primitive (1), then when the equation is solved for A, the function of x, y, to which it will be found equivalent, must contain radicals, since A has more than one value; hence the differential of this equation must contain the same radicals (*Diff. Calc.* p. 99), and, consequently, the expression for $\frac{dy}{dx}$, derived from it will have as many values as there are units in the highest exponent of A in the primitive (1). But if, instead of getting rid of A by the above process, we differentiate the primitive, and then eliminate the term containing A, and afterwards, by means of this result and the preceding, eliminate that containing A^2, and so on, we shall in this way introduce no radicals, and shall yet finally obtain, as before, a differential equation without A: it follows, therefore, that as $\frac{dy}{dx}$ must necessarily have the same values here as in the former equation, $\frac{dy}{dx}$ must enter in the same power that the constant A has in the primitive, and that, by solving the equation for $\frac{dy}{dx}$, we shall obtain the same expression as by the former process. Let us now briefly examine the differential equations of the succeeding orders.

The differential equation $\frac{d^2u}{dx^2} = 0$, immediately derived from the primitive, contains two constants fewer than that primitive; but, by elimination, a differential equation of the second order may be obtained, in which any two that may be proposed of the constants in the original primitive shall be absent, and there are two distinct ways in which this elimination may be performed. Thus, if A and B are the constants to be absent from the equation of the second order, then let us take the two equations of the first order, in the one of which A enters but not B, and in the other B but not A; from the first and its differential eliminate A, from the second and its differential eliminate B, and we shall have in each case the required differential equation of the second order. This equation has, therefore, two distinct primitives of the first order although but one final integral. As any two of the constants in the original primitive may be eliminated in this way, it follows that there are altogether as many differential equations of the second order as the n constants admit of combinations two and two, that is, there are $\dfrac{n\,(n-1)}{1\cdot 2}$ such equations derivable from the primitive.

In like manner, by elimination, we may obtain a differential equation of the third order, in which any three proposed constants of the original integral shall be absent, and there are three ways in which the equations of the second order enable us to do this; for, let there be taken those three of this order in which are absent the constants A, B from the first, A, C from the second, and B, C from the third; then, eliminating C from the first and its differential, eliminating B from the second and its differential, and A from the third and its differential, we obtain in each case a differential equation of the third order, without A, B, C; this equation, therefore, has *three* primitives of the second order; also, as the number of combinations of n things by threes is $\dfrac{n\,(n-1\,(n-2)}{1\cdot 2\cdot 3}$, it follows that this is the number of equations of the third order derivable from the primitive. Without pursuing this reasoning further we may obviously conclude that any of the equations of the mth order has m primitives of the $m-1$th order which are all different, and that the total number of equations of the mth order, derivable from the original primitive, is expressed by the number of different ways in which m of the n constants of the original primitive can be combined; the number of equations of the mth order is, therefore,

$$\frac{n\,(n-1(\,(n-2)\ldots(n-m+1)}{1\cdot 2\cdot 3\ldots\ldots m}.$$

If from any two of these the mth differential coefficient, or that which marks its order, be eliminated the result must necessarily be one of the m primitives of the $m - 1$th order; in like manner, by eliminating from any two differential equations of the same order the coefficient which marks that order, we shall arrive at one of the primitives of the preceding order, and thus at length at the final integral.

In the preceding discussion we have considered $F(x, y) = 0$ to be the complete nth integral of $\dfrac{d^n u}{dx^n}\, dx^n = 0$, containing all the n arbitrary constants, of which one is introduced at every integration; and it has also been seen that in setting out from this complete primitive, the equation $\dfrac{d^n u}{dx^n} = 0$ may be obtained from n different equations of the preceding or $n - 1$th order: from one by direct differentiation, and from the others by differentiation and elimination combined; hence, conversely setting out from $\dfrac{d^n u}{dx^n} = 0$, there must exist, besides that integral of the preceding order given immediately by integration, $n - 1$ other equations of the same order that are equally integrals of the proposed. It might not be amiss to call these *indirect* integrals, and the other the *direct* integral. If a differential equation of the first order be proposed for integration, we may by differentiation deduce from it all the succeeding direct differential equations, to the mth of which the preceding or $m - 1$th will be the direct integral; if, therefore, we can by any means obtain one of the indirect integrals, we shall then have altogether m equations preceding the mth differential, and in which there will enter $m - 1$ differential coefficients; all these may therefore be eliminated, and thus a final equation in x and y obtained, which will be the complete integral of the proposed.

Lastly, since direct differentiation introduces no powers of the differential coefficients, it follows that if we have a differential equation containing powers of the coefficients, we may be sure that it is one of the indirect differential equations derived from the primitive.

The *degree* of a differential equation is determined by the highest power of that differential coefficient which marks its order when the equation is freed from radicals.

(83.) It may not be amiss to confirm and illustrate the foregoing theory by an example.

Let us take the primitive equation of the first degree

$$y + ax + b = 0 \ . \ . \ . \ . \ (1),$$

then, by differentiation, we get the direct differential equation of the first order:

$$\frac{dy}{dx} + a = 0 \ . \ . \ . \ . \ (2),$$

wherein the constant b has disappeared.

By eliminating the constant a, still common to both these equations, we get the indirect differential equation of the first order:

$$y - x\frac{dy}{dx} + b = 0 \ . \ . \ . \ . \ (3).$$

The equation (1) is the common primitive of both these, b being the arbitrary constant when it is considered as the integral of (2), and a the arbitrary constant when considered as the integral of (3). The direct differential equation of the second order as derived from (2) is

$$\frac{d^2 y}{dx^2} = 0,$$

the indirect equation derived by eliminating b from (3), and its differential is also $\frac{d^2y}{dx^2} = 0$, the two equations (2), (3), being the two first primitives of this equation of the second order.

Let the primitive be

$$ax^2 - y + bx + c = 0.$$

By differentiating we have

$$2\,ax - \frac{dy}{dx} + b = 0 \ . \ . \ . \ . \ (1).$$

Eliminating from these equations first a and then b we have the two indirect differential equations

$$x\,\frac{dy}{dx} - 2y + bx + 2c = 0,\ ax^2 + y - x\frac{dy}{dx} - c = 0 \ \ldots\ (2),$$

and from each of these equations a different one of the three arbitrary constants a, b, c, has disappeared. Now let us seek that particular equation of the third order in which the constants b and c shall be absent; this will be had by differentiating (1), which gives

$$2a - \frac{d^2 y}{dx^2} = 0,$$

or by differentiating the second of (2), which gives

$$2ax - x\,\frac{d^2 y}{dx^2} = 0,\ \text{or } 2a - \frac{d^2 y}{dx^2} = 0$$

the same as before; hence (1) and the second of (2) are the two first primitives of this differential equation of the second order.

CHAPTER III.

ON THE INTEGRATION OF DIFFERENTIAL EQUATIONS OF TWO VARIABLES, AND OF THE FIRST ORDER AND DEGREE.

(84.) A differential equation, whatever be the number of variables it may contain, may obviously always be integrated whenever we can either separate the variables or render it an exact differential. In the former case the integration will be reduced to that of a series of differentials of one variable, and in the latter case the integration is effected by the method pointed out in last chapter. We shall here inquire how equations of the first order and degree containing two variables may be thus prepared for integration.

Separation of the Variables.

(85.) We shall first consider the general form

$$\mathrm{X}\, dy + \mathrm{Y}\, dx = 0,$$

which is the simplest for which the variables are separable: X being a function of x without y, and Y a function of y without x.

Dividing this equation by XY, the product of the coefficients, it becomes

$$\frac{dy}{\mathrm{Y}} + \frac{dx}{\mathrm{X}} = 0,$$

an equation in which the variables are separated, therefore

$$\int \frac{dy}{\mathrm{Y}} + \int \frac{dx}{\mathrm{X}} = 0$$

is the sought integral or equation between x and y.

As a particular example let the equation $(1 + x^2)\, dy = y^{\frac{1}{2}}\, dx$ be proposed.

Dividing by the product of the coefficients, the equation becomes

$$\frac{dy}{y^{\frac{1}{2}}} = \frac{dx}{1 + x^2};$$

hence, by integration, we have

$$2y^{\frac{1}{2}} = \tan.^{-1} x + C$$

the required relation between x and y.

(86.) The following is another general form in which the variables are readily separable, viz.

$$XY\, dy + X'Y'\, dx = 0.$$

For dividing by XY' the equation becomes

$$\frac{Y\, dy}{Y'} + \frac{X'\, dx}{X} = 0,$$

where the first coefficient is a function y without x, and the second a function of x without y.

As an example let the equation be

$$x^2 y\, dx + (3y + 1)\, x^{\frac{3}{2}}\, dy = 0,$$

which belongs to the above form, since the coefficient of each differential is the product of a function of x by a function of y. Dividing by $y\, x^{\frac{3}{2}}$, it becomes

$$x^{\frac{1}{2}}\, dx + \frac{3y + 1}{y}\, dy = 0,$$

and taking the integrals, we have

$$\frac{2}{3}\, x^{\frac{3}{2}} + 3y + \log. y = C$$

the required relation between x and y.

(87) In the form

$$dy + Xy\, dx = X'\, dx \quad \ldots \quad (1),$$

called a *linear equation*, because y enters only in the first power, the variables may be separated.

To effect this separation put $z = \dfrac{y}{X}$, X, being any arbitrary function of x, then

$$y = zX, \therefore dy = zdX, + X,dz,$$

and the proposed will become

$$zdX, + X, (dz + Xz\,dx) = X'dx.$$

Now if the arbitrary function X' be determined from the condition

$$zdX, = X'dx, \text{ then } X, (dz + Xz\,dx) = 0 \quad \ldots \ldots \quad (2),$$

from which last equation we get

$$\frac{dz}{z} = - Xdx \therefore \log. z = -\int Xdx,$$

that is passing from logarithms to the numbers, these all being powers of the base e whose exponents are the logarithms

$$z = e^{-\int Xdx},$$

Substituting this value of z in the first of the equations (2) we get

$$dX, = \frac{X'dx}{z} = X'e^{\int Xdx}\,dx$$

$$\therefore X, = \int X' e^{\int Xdx}\,dx,$$

but we assumed $y = zX,$; hence, by substitution,

$$y = e^{-\int Xdx} \left\{ \int X' e^{\int Xdx}\,dx \right\} \quad \ldots \ldots \quad (3),$$

which expresses the relation between x and y.

To this form may be reduced the more general form

$$dy + Xy\,dx = X'y^{n+1}\,dx \quad \ldots \ldots \quad (4).$$

For substituting

$$z \text{ for } \frac{1}{y^n} \therefore y = \frac{1}{z^{\frac{1}{n}}} \therefore dy = - \frac{dz}{nz^{\frac{1}{n}+1}}$$

the equation becomes

$$- \frac{dz}{nz^{\frac{1}{n}+1}} + X \frac{dx}{z^{\frac{1}{n}}} = X' \frac{dx}{z^{\frac{1}{n}+1}},$$

or multiplying by $-n z^{-\frac{1}{n}+1}$,

$$dz - nXz\,dx = -nX'\,dx,$$

which agrees with the form (1), and consequently, by (3),

$$z = \frac{1}{yn} = e^{n\int Xdx}\left\{-n\int X'e^{-n\int Xdx}\,dx\right\} \ldots (5).$$

We shall add an example or two of the foregoing forms.

EXAMPLES.

1. Given

$$dy + y\,dx = ax^3\,dx$$

to determine the relation between x and y.

In this example $X = 1$, $X' = ax^3$,

$$\therefore \int Xdx = x^* \quad \therefore \int X'e^{\int Xdx}\,dx = a\int x^3 e^x\,dx = (\text{ex. 2, p. 59})$$

$$ae^x (x^3 - 3x^2 + 6x - 6) + C$$

$$\therefore y = a(x^3 - 3x^2 + 6x - 6) + Ce^{-x},$$

which is the relation between x and y required.

2. Given

$$(1 + x^2)\,dy - yx\,dx = a\,dx$$

to determine the relation between x and y.

Here

$$X = -\frac{x}{1 + x^2}, \quad X' = \frac{a}{1 + x^2}$$

$$\therefore \int Xdx = -\log.\sqrt{1 + x^2} \quad \therefore e^{\int Xdx} = \frac{1}{e^{\log.\sqrt{1+x^2}}},$$

but by logarithms

---

* It is useless to add a constant to the integral of $X\,dx$, for suppose we do this and write thus the integral so completed $P + c$, then the general formula (3) may be written

$$y = e^{-P} \cdot e^{-c}\left\{\int X' e^{P} \cdot e^{c}\,dx\right\}$$

$$= e^{-P}\left\{\int X' e^{P}\,dx\right\}.$$

$$\sqrt{1+x^2} = e^{\log.\sqrt{1+x^2}} \quad \therefore e^{\int X dx} = \frac{1}{\sqrt{1+x^2}}$$

$$\therefore \int X' e^{\int X dx} dx = \int \frac{a\,dx}{(1+x^2)^{\frac{3}{2}}} = \frac{ax}{\sqrt{1+x^2}} + C \text{ (see page 50,)}$$

$$\therefore y = ax + C\sqrt{1+x^2},$$

the relation required.

3. Given $dy + y\,dx = xy^3\,dx$ to determine the relation between x and y.

This equation comes under the form (4), and we have here

$$X = 1, \; X' = x, \; n = 2$$

$$\therefore \int X dx = x \; \therefore \; \int X' e^{-2\int X dx} dx = \int x e^{-2x} dx = x e^{-2x} + \frac{1}{2} e^{-2x} + C.$$

see ex. 5, p. 59; hence, by the formula (5),

$$\frac{1}{y^2} = Ce^{2x} + x + \frac{1}{2},$$

the relation required.

(88.) The summation of some extensive classes of infinite series may be made to depend on the integration of linear differential equations; we shall here give an instance or two.

4. Required the sum of the infinite series:

$$y = \frac{x}{1} + \frac{x^3}{1 \cdot 3} + \frac{x^5}{1 \cdot 3 \cdot 5} + \frac{x^7}{1 \cdot 3 \cdot 5 \cdot 7} + \&c.$$

By differentiating

$$\frac{dy}{dx} = 1 + \frac{x^2}{1} + \frac{x^4}{1 \cdot 3} + \frac{x^6}{1 \cdot 3 \cdot 5} + \&c.$$

Transposing the first term and dividing by x,

$$\frac{dy}{x\,dx} - \frac{1}{x} = \frac{x}{1} + \frac{x^3}{1 \cdot 3} + \frac{x^5}{1 \cdot 3 \cdot 5} + \&c.,$$

which is the proposed series; consequently,

$$\frac{dy}{x\,dx} - \frac{1}{x} = y \;\therefore\; dy - xy\,dx = dx,$$

and thus the sum of the series depends upon the integration of the linear differential equation

$$dy - xy\,dx = dx,$$

in which

$$X = -x,\; X' = 1 \;\therefore\; \int X\,dx = -\frac{1}{2}\,x^2$$

$$\therefore \int X' \epsilon^{\int X dx}\,dx = \int e^{-\frac{1}{2}x^2}\,dx \;\therefore\; y = e^{\frac{1}{2}x^2}\left\{\int e^{-\frac{1}{2}x^2}\,dx\right\}.$$

This, although an analytical expression for the sum of the series, does not, however, enable us to exhibit that sum in finite terms, because the integral within the brackets can be expressed only by series. The proposed series will, however, at once give us the development of this integral, or of the integral $\int e^{-z^2}\,dz$, by putting $z\sqrt{2}$ for x, for we shall then have

$$\int e^{-z^2}\,dz = \frac{ye^{-z^2}}{\sqrt{2}} = e^{-z^2}\left\{\frac{z}{1} + \frac{2z^3}{1\cdot 2} + \frac{4z^5}{1\cdot 3\cdot 5} + \right.$$

$$\left. \frac{8z^7}{1\cdot 3\cdot 5\cdot 7} + \&c.\right\}.$$

5. Required the sum of the infinite series

$$y = 1 + \frac{m}{n}\,x + \frac{m(m+1)}{n(n+1)}\,x^2 + \frac{m(m+1)(m+2)}{n(n+1)(n+2)}\,x^3 + \&c.$$

Multiplying by x^{n-1},

$$yx^{n-1} = x^{n-1} + \frac{m}{n}\,x^n + \frac{m(m+1)}{n(n+1)}\,x^{n+1} + \&c.$$

Differentiating,

$$x^{n-1}\frac{dy}{dx} + (n-1)\,yx^{n-2} = (n-1)\,x^{n-2} + mx^{n-1} +$$

$$\frac{m(m+1)}{n}\,x^n + \&c.$$

Multiplying this by x^{m-n},

$$x^{m-1}\frac{dy}{dx} + (n-1)\,yx^{m-2} = (n-1)\,x^{m-2} + mx^{m-1} +$$

$$\frac{m(m+1)}{n} x^m + \&c.$$

Multiplying now by dx, and integrating,

$$\int \left(x^{m-1} \frac{dy}{dx} + (n-1) yx^{m-2} \right) dx = \frac{n-1}{m-1} x^{m-1} + x^m + \frac{m}{n} x^{m+1} +$$

$$\frac{m(m+1)}{n(n+1)} x^{m+2} + \&c.$$

$$= \frac{n-1}{m-1} x^{m-1} + x^m y.$$

Therefore, by differentiating

$$x^{m-1} \frac{dy}{dx} + (n-1) yx^{m-2} = (n-1) x^{m-2} + x^m \frac{dy}{dx} + mx^{m-1} y$$

$$\therefore dy + \frac{n-1-mx}{x(1-x)} ydx = \frac{(n-1) dx}{x(1-x)},$$

which is a linear differential equation agreeing with the form (1), and in which

$$X = \frac{n-1-mx}{X(1-x)}, \; X' = \frac{n-1}{x(1-x)}$$

$$\therefore \int X dx = (n-1) \int \frac{dx}{x(1-x)} - m \int \frac{dx}{1-x}$$

$$= (n-1) \int \frac{dx}{x} + (n-1) \int \frac{dx}{1-x} - m \int \frac{dx}{1-x} =$$

$$(n-1) \log. x - (n-1) \log. (1-x) +$$

$$m \log. (1-x) = \frac{\log. x^{n-1}}{\log. (1-x)^{n-m-1}} \therefore e^{\int X dx} = \frac{x^{n-1}}{(1-x)^{n-m-1}},$$

consequently

$$y = \frac{(1-x)^{n-m-1}}{x^{n-1}} \left\{ (n-1) \int \frac{x^{n-2} dx}{(1-x)^{n-m}} \right\},$$

the integral within the brackets being that of a rational fraction; it must be corrected, so that y may be equal to 1, when $x = 0$.

    6. Given

$$dy + ydx = ax^n dx$$

to determine the relation between x and y.

$$y = Ce^{-x} + x^n - nx^{n-1} + n(n-1)x^{n-2} - \&c.$$

7. Given

$$dy - \frac{a}{1-x}\,ydx = bdx,$$

to determine the relation between x and y.

$$y = \frac{C}{(1-x)^a} - \frac{b}{a+1}(1-x).$$

8. Required the sum of the infinite series

$$y = 1 + \frac{m}{n}x + \frac{m+1}{n+1}x^2 + \frac{m+2}{n+2}x^3 + \&c.$$

$$y = x^{1-n}\left\{x^{n-1} + m\int\frac{x^{n-1}\,dx}{1-x} + \int\frac{x^n\,dx}{(1-x)^2}\right\}.$$

9. Given

$$dy + \frac{xy\,dx}{1-x^2} = xy^{\frac{1}{2}}\,dx$$

to determine the relation between x and y.

$$y^{\frac{1}{2}} = C\,(1-x^2)^{\frac{1}{4}} - \frac{1}{3}(1-x^2).$$

(88.) In addition to the foregoing general forms, the separation of the variables may always be effected in differential equations of the first order containing two variables, whenever they are homogeneous.

In order to establish this, we must first prove that if u be a homogeneous function of x and y, the degree of homogeneity being n, then z may always be put under the form

$$u = x^n\,F\left(\frac{x}{y}\right).$$

For suppose the homogeneous function to be developed into a series of monomials, such as

$$Ax^p y^q,\ Bx^{p'} y^{q'},\ Cx^{p''} y^{q''},\ \&c.$$

then, by virtue of the homogeneity,

$$p + q = n;\ p' + q' = n,\ p'' + q'' = n,\ \&c.$$

If now we divide each of the terms by x^n, the first will become

$$\frac{A x^p y^q}{x^n} = \frac{A y^q}{x^{n-p}} = \frac{A y^q}{x^q} = A \left(\frac{y}{x}\right)^q,$$

and the others will become

$$B \left(\frac{y}{x}\right)^{q'}, \; C \left(\frac{y}{x}\right)^{q''}, \; \&c.$$

so that we shall have

$$\frac{u}{x^n} = F\left(\frac{y}{x}\right) \therefore u = x^n F\left(\frac{y}{x}\right).$$

Let now the equation

$$P dx + Q dy = 0$$

be proposed, in which P and Q are homogeneous functions of x and y of the degree n.

Then, dividing by x^n, it takes the form

$$F\left(\frac{y}{x}\right) dx + f\left(\frac{y}{x}\right) dy = 0,$$

or, substituting z for $\frac{y}{x}$,

$$Fz dz + fz dy = 0 \text{ or } Fz + fz \frac{dy}{dx} = 0.$$

In order to eliminate dy, let us differentiate the equation $\frac{y}{x} = z$, or, rather, $y = z x$, and we get

$$\frac{dy}{dx} = z + \frac{x dz}{dx},$$

which, substituted in the foregoing equation, reduces it to

$$Fz + fz \left(z + \frac{x dz}{dx}\right) = 0$$

$$\therefore \frac{x dz}{dx} = -\frac{Fz + z fz}{fz}$$

$$\therefore \frac{dx}{x} = -\frac{dz fz}{fz + z fz}$$

an equation involving the same relation between x and y as the proposed, and in which the variables x, z are separated, and which gives, by integration,

$$\log. x = -\int \frac{dz fz}{Fz + z fz}, \cdot$$

so that, after having obtained the integral on the right, we shall only have to substitute in the result $\frac{y}{x}$ for z, and we shall then have the relation between x and y sought. Let us apply this general process to one or two particular cases.

EXAMPLES.

1. To determine the relation between x and y in the equation

$$\frac{x^2 + yx}{x - y} dy = y dx.$$

Multiplying by $x - y$, this becomes

$$(x^2 + yx)\, dy = (xy - y)\, dx,$$

which is homogeneous, and of two dimensions, therefore, dividing by x^2, we have

$$(1 + \frac{y}{x})\, dy = (\frac{y}{x} - \frac{y^2}{x^2})\, dx,$$

that is, substituting z for $\frac{y}{x}$,

$$(1 + z)\frac{dy}{dx} = (z - z^2),$$

but, since

$$y = zx \therefore \frac{dy}{dx} = z + \frac{x dz}{dx};$$

hence, by substitution and division,

$$z + \frac{x dz}{dx} = \frac{z(1 - z)}{1 + z}$$

$$\therefore \frac{xdz}{dx} = -\frac{2z^2}{1+z}$$

$$\therefore \frac{dx}{x} = -\frac{1+z}{2z^2}\, dz,$$

consequently

$$\log. x = -\frac{1}{2}\int \frac{dz}{z^2} - \frac{1}{2}\int \frac{dz}{z} = \frac{1}{2z} - \frac{1}{2}\log. z + C =$$

$$\frac{x}{2y} + \frac{1}{2}\log. \frac{y}{x} + C,$$

which is the relation required.

2. Given

$$x\, dy - y\, dx = dx\, \sqrt{x^2 - y^2}$$

to determine the relation between x and y.

Dividing by x, this equation becomes

$$dy - \frac{y}{x}\, dx = dx\, \sqrt{1 - \frac{y^2}{x^2}},$$

or, substituting z for $\frac{y}{x}$,

$$\frac{dy}{dx} - z = \sqrt{1 - z^2},$$

or, putting instead of $\frac{dy}{dx}$, its value

$$\frac{dy}{dx} = z + \frac{xdz}{dx},$$

it becomes

$$\frac{xdz}{dx} = \sqrt{1 - z^2}$$

$$\therefore \frac{dx}{x} = \frac{dz}{\sqrt{1 - z^2}}$$

$$\therefore \log. x = \sin.^{-1} z + C = \sin.^{-1} \frac{y}{x} + C,$$

the relation required.

It ought to be remarked, that if any function of $\frac{y}{x}$, unmixed with the variables in any other form, enter an equation which would otherwise be homogeneous, the equation may be treated as if it were homogeneous, and the relation between x and y will be determined; the following example will illustrate this remark:

3. Given the equation

$$xy\, dy - y^2\, dx = (x+y)^2 e^{-\frac{y}{x}}\, dx$$

to determine the relation between x and y.

Dividing by x^2, and substituting z for $\frac{y}{x}$, we have

$$z\frac{dy}{dx} = z^2 + (1+z)^2 e^{-z},$$

that is, since

$$\frac{dy}{dx} = z + \frac{x\,dz}{dx}$$

$$\frac{x}{dx} \cdot z\,dz = (1+z^2)e^{-z}$$

$$\therefore \frac{dx}{x} = \frac{e^z z\,dz}{(1+z)^2}$$

$$\therefore \log. x = \frac{e^z}{1+z} - \frac{e^{\frac{y}{x}}}{1+\frac{y}{x}} + C = \frac{xe^{\frac{y}{x}}}{x+y} + C.$$

4. Given

$$xdy - ydx = dx\sqrt{x^2 + y^2}$$

to determine the relation between x and y.

$$x^2 = 2Cy + C^2.$$

5. Given

$$ydy + (x+2y)\, dx = 0$$

to determine the relation between x and y.

$$\log. (x+y) + \frac{x}{x+y} = C.$$

6. Given

$$y^2 \, dy = 3yx \, dx - x^2 \, dy$$

to determine the relation between x and y.

$$y^2 - 2x^2 = Cy^{\frac{2}{3}}.$$

89. Equations may sometimes be rendered homogeneous by means of certain substitutions. Thus the general example

$$(mx + ny + p) \, dx + (ax + by + c) \, dy = 0$$

will become homogeneous, if we put

$$x = x' + a, \; y = y' + \beta,$$

and then determine a and β from the conditions

$$p + ma + n\beta = 0, \; c + aa + b\beta = 0,$$

for it will then become

$$(mx' + ny' + ny') \, dx' + (ax' + by') \, dy' = 0.$$

(90.) Let us now examine the different integrable cases of

The Equation of Riccati.

$$dy + by^2 \, dx = ax^m \, dx \; \ldots \ldots \; (A).$$

1. Let $m = 0$, then this equation becomes

$$dy + by^2 \, dx = a \, dx \; \therefore \; dx = \frac{dy}{a - by^2},$$

where the variables are separated, and, by integrating, we have for the relation between x and y,

$$2a^{\frac{1}{2}} x + C = \log. \; (a^{\frac{1}{2}} + b^{\frac{1}{2}} y) - \log. (a^{\frac{1}{2}} - b^{\frac{1}{2}} y).$$

2. When m is not 0, let us inquire for what values of m the equation may be rendered homogeneous. For this purpose put $y = z^k$, and it becomes

$$kz^{k-1} \, dz + bz^{2k} \, dx = ax^m \, dx,$$

which cannot be homogeneous, unless

$$k - 1 = 2k = m \therefore k = -1 \therefore m = -2;$$

hence the only case of Riccati's equation that can be rendered homogeneous is

$$dy + by^2\, dx = ax^{-2}\, dx,$$

which is rendered so by the substitution of $z-1$ for y.

3. Besides the case 1, above, which is the simplest for which the variables are separable, *Euler* has found an infinite number of other cases, in which the separation is possible. To discover these, let us assume

$$y = -y'\, x'^2 + \frac{x'}{b} \ldots \ldots (1),$$

either of the variables x', y' being arbitrary, then

$$dy = -2x'y'\, dx' - x'^2\, dy' + \frac{dx'}{b} \ldots \ldots (2),$$

and

$$y^2 = y'^2\, x'^4 - \frac{2y'\, x'^3}{b} + \frac{x'^2}{b} \ldots \ldots (3).$$

Now, by adding to the first of these equations the second multiplied by bdx, the sum of their second members will be equal to the first member of the proposed equation (A). Let us see, therefore, whether we cannot assume for the arbitrary quantity x' such a value as may render the result of this addition similar in form to the first member of (A), as well as equivalent to it in value. The suitable expression for x' is easily perceived to be

$$x' = \frac{1}{x} \therefore x = \frac{1}{x'} \therefore bdx = -\frac{bdx'}{x'^2} \ldots \ldots (4),$$

for, multiplying (3) by this last, and adding the product to (2), we shall have

$$-x'^2\, dy' - by'^2\, x'^2\, dx' = -ax'^{-m-2}\, dx',$$

or

$$dy' + by'^2\, dx' = ax'^{-m-4}\, dx' \ldots \ldots (5),$$

which equation is similar to the proposed, and has the same coefficients a, b with the same signs. By means of this transformation it appears from the first case that the proposed is integrable when $-m-4 = 0$, that is, when $m = -4$, and generally that for whatever value n of m the proposed becomes integrable, it also becomes integrable for another value, viz. that given by

$$- m - 4 = n \therefore m = - n - 4 \ldots \ldots (6).$$

It appears, therefore, from what has now been said, that whenever we have any case of Riccati's equation that may be compared to the form (5), it may be immediately changed to the form (A), the variables in the transformed equation being related to those in the proposed as in (1), (4).   That is, the equation

$$dy + by^2 \, dx = ax^{-m-4} \, dx \ldots \ldots (7)$$

is the same as

$$dy' + by'^2 \, dx' = ax'^m \, dx' \ldots \ldots (8),$$

in which

$$x' = \frac{1}{x} \text{ and } y' = - yx^2 + \frac{x'}{b}.$$

Again, assume in the equation (A),

$$y = \pm \frac{1}{y'}, \therefore dy = \pm \frac{dy'}{y'^2} \ldots \ldots (9),$$

and it becomes

$$\mp \, dy' + b dx = a y'^2 \, x^m \, dx,$$

in which, if we put

$$x^{m+1} = x' \therefore x^m \, dx = \frac{dx'}{m + 1} \ldots \ldots (10)$$

we have the transformed equation

$$\mp \, dy' + \frac{b}{m + 1} x'^{-\frac{m}{m+1}} = \frac{a}{m + 1} y'^2 \, dx',$$

or

$$dy' \pm \frac{a}{m + 1} y'^2 \, dx' = \pm \frac{b}{m + 1} x'^{-\frac{m}{m+1}} \ldots \ldots (11).$$

Now it may be remarked here that the only integrable cases of Riccati's equation hitherto discovered, and which we are about to exhibit, are those in which the exponent m is negative, and numerically greater than unity, with the exception of the *fundamental case*, $m = 0$, already considered; therefore, since, in the equation (11), $m + 1$ is necessarily negative, it follows that when the upper sign has place, that is, when $\dfrac{1}{y'}$ is put for y in (A), then in the transformed equation (11), the signs of the coefficients, in the first and second members, will be respectively opposite to those of the coefficients in the second and first members of (A). But when the lower sign has place in (11), that is, when $-\dfrac{1}{y'}$ is put for y, then the coefficients in the first and second members of (11) have respectively the same signs as those in the second and first members of (A). Hence it follows that whatever be the signs of a, b, we may always infer that the equation

$$dy' \pm \frac{a}{m+1} y'^2 dx' = \pm \frac{b}{m+1} x'^{-\frac{m}{m+1}} \ldots \ldots (12)$$

is the same as

$$dy'' + by''^2 dx'' = ax''^m dx'' \ldots \ldots (13),$$

in which

$$x'' = x'^{m+1} \text{ and } y'' = \pm \frac{1}{y'},$$

abstraction being made of the signs of a and b.

It appears, therefore, that whenever (12) is integrable, (13) is also integrable, and we may now shew that there are an infinite number of integrable cases of Riccati's equation. For, let n be one of the integrable cases of (12), then

$$-\frac{m}{m+1} = n \therefore m = -\frac{n}{n+1} \ldots \ldots (14),$$

so that by (13) this value of m belongs to an integrable case.

The integrable cases hitherto found are those where the exponent is either 0 or -4, the first put for n in this formula gives no new case; but substituting the latter we have

$$m = -\frac{-4}{-4+1} = -\frac{4}{3},$$

which is, therefore, an integrable case; hence, putting this value for n in (6), we have another integrable case, viz.

$$m = \frac{4}{3} - 4 = -\frac{8}{3}.$$

Putting this for n in (14), we have another integrable case, viz.

$$m = -\frac{-\frac{8}{3}}{-\frac{8}{3}+1} = -\frac{8}{5},$$

and thus, by using alternately the expressions for m in (6) and (14), we find that besides the two cases in which the values of m are 0, and -2 the equation is also integrable, when m is any term in the infinite series

$$-4, -\frac{4}{3}, -\frac{8}{3}, -\frac{8}{5}, -\frac{12}{5}, -\frac{12}{7}, -\frac{16}{7}, \&c. \ldots \ldots (15),$$

the general term being the negative number

$$m = -\frac{4q}{2q \pm 1},$$

which we may consider as the *criterion of integrability of Riccati's equation*, when m is neither 0 nor -2, q being any number in the series 1, 2, 3, &c.

It must be observed that the first, third, fifth, and all the odd terms of the series (15) arise from the formula (7), and the second, fourth, sixth, and all the even terms, from the formula (12), so that, when, in any proposed case, m is an odd term of the series (15), we must compare it with (7), and deduce the transformed equation (8), the exponent in which will be the preceding term in the series (15). We shall then have to compare this reduced equation with (12), and deduce the transformation (13) in which the exponent will be the next preceding term of (15), we shall, therefore, return with this equation to (7), and so on, alternately using the forms (7) and (12) till at length the exponent in (7) becomes -4, when the transformation (8) will be the final equation, and will be immediately integrable. An example will clearly illustrate this:

EXAMPLES.

(91.) 1. To determine the relation between x and y in the equation

$$dy + y^2\, dx = \frac{2dx}{x^{\frac{8}{3}}}.$$

Since the exponent $m = -\dfrac{8}{3}$ is found among the terms of the series (15), we are sure that the proposed is an integrable form, and our object is to reduce it by successive transformations to the fundamental form, $m = 0$. In order to this, as $-\dfrac{8}{3}$ is an odd term of the series we must commence these transformations by comparing the proposed with (7), so that

$$a = 2,\ m + 4 = \frac{8}{3} \ \therefore\ m = -\frac{4}{3} \ \text{and}\ b = 1,$$

and the first transformation (8) is

$$dy' + by'^2\, dx' - 2x'^{-\frac{4}{3}} \ \ldots\ .(1);$$

comparing this with (12), we have

$$\frac{a}{m+1} = 1,\ \frac{b}{m+1} = 2 \ \text{and}\ \frac{m}{m+1} = \frac{4}{3}$$

from which we get

$$a = -3,\ b = -6,\ m = -4,$$

so that the second transformation (13) is

$$dy'' - 6y''^2\, dx'' = -3x''^{-4}\, dx'' \ \ldots\ . (2).$$

Returning with this to equation (7), we have for the final transformation (8)

$$dy''' - 6y'''^2\, dx''' = -3dx''' \ \ldots\ . (3),$$

of which the primitive is

$$x''' = \frac{1}{6} \int \frac{dy''}{y'''^2 - \frac{1}{2}} = -\frac{1}{6\sqrt{2}} \log. C \frac{y''' + \sqrt{\frac{1}{2}}}{y''' - \sqrt{\frac{1}{2}}} \ \ldots\ . (4).$$

It remains then to substitute in this equation, for x''', y''', their proper values in terms of x, y. From (4) and (1) we have

$$x' = \frac{1}{x}, \; y' = -\, yx^2 + x,$$

also

$$x''' = \frac{1}{x''}, \; y''' = -\, y'x''^2 - \frac{x''}{6},$$

and from (9) and (10)

$$x'' = x^{\frac{1}{3}}, \; y'' = \frac{1}{y'} = \frac{1}{x - yx^2}$$

$$\therefore x''' = x^{-\frac{1}{3}}, \; y''' = -\, \frac{6 + x^{\frac{2}{3}}(1 - yx)}{6x^{\frac{1}{3}}(1 - yx)};$$

hence, by substituting these values in the equation (4), we have the following equation between x and y, viz.

$$x^{-\frac{1}{3}} = -\, \frac{1}{6\sqrt{2}} \; \log. \; C \, \frac{6 - x^{\frac{1}{3}}(3\sqrt{2} - x^{\frac{1}{3}})(1 - yx)}{6 + x^{\frac{1}{3}}(3\sqrt{2} + x^{\frac{1}{3}})(1 - yx)},$$

or, since

$$6\sqrt{2}x^{-\frac{1}{3}} = 6\sqrt{2}x^{-\frac{1}{3}} \log. \, e = \log. \, e^{6\sqrt{2}x^{-\frac{1}{3}}},$$

the equation may be written thus, by passing from the logarithms to the numbers

$$e^{6\sqrt{2}x^{-\frac{1}{3}}} \left\{ \frac{6 + x^{\frac{1}{3}}(3\sqrt{2} + x^{\frac{1}{3}})(1 - yx)}{6 - x^{\frac{1}{3}}(3\sqrt{2} - x^{\frac{1}{3}})(1 - yx)} \right\} = C.$$

2. To determine the relation between x and y, in the equation

$$dy + y^2 \, dx = \frac{a^2 \, dx}{x^4}$$

$$e^{\frac{2a}{x}} \left\{ \frac{x(xy - 1) + a}{x(xy - 1) - a} \right\} = C.$$

3. To determine the relation between x and y, in the equation

$$dy + y^2 \, dx = -\, a^2 \, x^{-4}$$

$$\frac{a}{x} + C = \tan.^{-1} \frac{x^2 y - x}{a}.$$

4. To determine the relation between x and y, in the equation

$$dy = (y^2 + 2x^{-\frac{8}{3}})\, dx$$

$$\frac{3\sqrt{2}}{x^{\frac{1}{3}}} = \tan.^{-1} \frac{x^{\frac{2}{3}} + yx^{\frac{5}{3}} - 6}{3\sqrt{2}x^{\frac{1}{3}}(xy + 1)} + C.$$

It may be observed before terminating this article, that every equation of the form

$$dy + by^2\, x^q\, dx = ax^p\, dx \;\; \ldots \ldots \; (B),$$

which indeed is the form in which the equation which bears this name was proposed by Riccati, may be reduced to the simpler form (A), by substituting

$$x^q\, dx \text{ for } dz,$$

for we thus get

$$\frac{x^{q+1}}{q+1} = z \text{ and } x = \{(q+1)\, z\}^{\frac{1}{q+1}},$$

and, differentiating this last equation, we have

$$dx = \frac{z^{\frac{-q}{q+1}}\, dz}{(q+1)^{\frac{q}{q+1}}},$$

also

$$x^p = \{(q+1)\, z\}^{\frac{p}{q+1}},$$

and these values, substituted in the equation (B) transform it to

$$dy + by^2\, dz = a\,(q+1)^{\frac{p-q}{q+1}}\; z^{\frac{p-q}{q+1}}\, dz,$$

which agrees with the form (A), page 202.

Having now considered the principal cases of differential equations of two variables, which may be integrated by the separation of the variables, it remains to examine those equations which may be converted into exact differentials.

On rendering Differential Equations exact.

(92.) Every differential equation

$$Mdx + Ndy = 0 \ldots (1)$$

necessarily implies a relation between x and y, which relation exhibits the primitive of that equation. It is not necessary, however, that (1) should arise from this primitive by direct differentiation, it may arise from eliminating a constant between this primitive and its direct differential, and if so it will not satisfy the condition of integrability (78) which has place only for direct differentials. The relation, however, between x and y is the same both in the direct and indirect differential equation, as already observed at page 183, and on this account it is easy to see that the one ought to become identical with the other, by introducing a factor, but that a factor will render every differential equation exact may be directly proved as follows:

Divide the equation (1) by Ndx and it becomes

$$p' + K = 0 \ldots (2),$$

K being put for $\dfrac{M}{N}$, and let us suppose that c is the constant, by the elimination of which from the primitive $F(x, y) = 0$ and its immediate differential the equation (1) or (2) has been produced. The same will be produced if we solve the primitive for c and differentiate the result; that is, putting the primitive under the form $c = f(x, y)$ and differentiating, we have

$$0 = Pp' + Q \therefore p' + \frac{Q}{P} = p' + K,$$

that is,

$$Pp' + Q = P(p' + K),$$

but the first member of this equation is the exact differential of $f(x, y)$; hence the second member is the exact differential of the same function, so that there always exists a factor P which will render any proposed differential (1) integrable.

Besides the factor P, there exists also an infinite number of others

that will render the proposed integrable, for representing the integral of
$MP\,dx + NP\,dy = 0$ by u, we shall have

$$du = MP dx + NP dy,$$

and multiplying each member by any arbitrary function of u, ϕu we get

$$\phi u\,du = \phi u\,(MP dx + NP dy),$$

and it is obvious that we may assume ϕu of an infinite number of
values that may render the first member of this last equation an exact
differential, and consequently the second member also.

As to the determination of one of these factors z, in the first instance
we know that since $Mz dx + Nz dy$ is an exact differential, we must
have the condition

$$\frac{d \cdot Mz}{dy} = \frac{d \cdot Nz}{dx},$$

that is,

$$\frac{M dz}{dy} + \frac{z dM}{dy} = \frac{N dz}{dx} + \frac{z dN}{dx}$$

$$\therefore z\left(\frac{dM}{dy} - \frac{dN}{dx}\right) = N\,\frac{dz}{dx} - M\,\frac{dz}{dy} \ldots \ldots (3),$$

from which equation we can deduce a value for z in particular circum-
stances, viz. 1st, when this factor happens to be a function of only one
of the variables, and 2d, if the differential expression is homogeneous.

(93.) Let z be a function of x only, then $\dfrac{dz}{dy} = 0$, and therefore
from equation (3) we deduce

$$\left(\frac{dM}{dy} - \frac{dN}{dx}\right)\frac{dx}{N} = \frac{dz}{z},$$

so that the first member cannot contain y; hence, if there exist a
factor z which is a function of x only, we must have in the first place
the condition

$$\left(\frac{dM}{dy} - \frac{dN}{dx}\right) \div N = Fx \ldots \ldots (4),$$

and then to determine z we have the equation

$$\log. z = \int Fx\,dx.$$

Let us take a particular case or two of this kind.

EXAMPLES.

1. To determine the primitive of the equation

$$ydx - xdy = 0.$$

Here

$$\left\{\frac{dM}{dy} - \frac{dN}{dx}\right\} \frac{1}{N} = \frac{2}{-x}$$

$$\therefore \log. z = -2\int \frac{dx}{x} = \log. \frac{1}{x^2}$$

$$\therefore z = \frac{1}{x^2},$$

therefore, multiplying the proposed by this, we have

$$\frac{ydx - xdy}{x^2} = 0,$$

of which the integral is

$$\frac{y}{x} = 0.$$

2. To determine the primitive of the equation

$$xdy + (b - 2y)\,dx = 0,$$

$$\left\{\frac{dM}{dy} - \frac{dN}{dx}\right\} \frac{1}{N} = -\frac{3}{x}$$

$$\therefore \log. z = -3\int \frac{dx}{x} = \log. \frac{1}{x^3}$$

$$\therefore z = \frac{1}{x^3},$$

therefore, multiplying the proposed by this,

$$\frac{dy}{x^2} + (b - 2y)\,\frac{dx}{x^3} = 0,$$

which is an exact differential, the primitive being

$$\frac{y}{x^2} - \frac{b}{2x^2} + C = 0.$$

3. Let the linear equation

$$dy + Pydx = Qdx$$

be proposed:

$$\left\{ \frac{d\mathrm{M}}{dy} - \frac{d\mathrm{N}}{dx} \right\} \frac{1}{\mathrm{N}} = \mathrm{P}$$

$$\therefore \ \log. z = \int P dx \ \therefore \ z = e^{\int P dx},$$

therefore, multiplying the proposed by this, we have

$$e^{\int P dx}\, dy + y\, e^{\int P dx}\, P dx = e^{\int P dx}\, Q dx,$$

of which the primitive is

$$e^{\int P dx}\, y = \int e^{\int P dx}\, Q dx$$

or

$$y = e^{-\int P dx} \left\{ \int e^{\int P dx}\, Q dx \right\}.$$

It must be observed that z is not necessarily the factor which will render the equation integrable, although the condition (4) have place, for there may not exist any such factor: we cannot affirm therefore that the proposed after having been multiplied by the factor z, as determined from that condition, has been rendered integrable till we have submitted it to the criterion of integrability (78). The following example from *Jephson's Fluxional Calculus* is not to be rendered integrable by any factor which is a function of x only, although the condition (4) has place.

4. Let the equation be

$$aydx + 2uxdy = xydx,$$

$$\left\{ \frac{d\mathrm{M}}{dy} - \frac{d\mathrm{N}}{dx} \right\} \frac{1}{\mathrm{N}} = \frac{1}{2ax}(a - x - 2a) = - \frac{a + x}{2ax}$$

$$\therefore \ \log. z = - \frac{1}{2}\int \frac{dx}{x} - \frac{1}{2a}\int dy = \frac{1}{\log. x^{\frac{1}{2}}} - \frac{x}{2a}$$

$$\therefore \ z = e^{\frac{1}{\log. \sqrt{x}}} \div e^{\frac{x}{2a}},$$

and if the proposed be multiplied by this the result will not satisfy the condition of integrability.

5. To determine the primitive of the equation

$$x^3\,dy + \left(4x^2\,y - \frac{1}{\sqrt{1 - x^2}}\right) dx = 0$$

$$x^4\,y + \sqrt{1 - x^2} = C.$$

6. To determine the primitive of the equation

$$ay\,dy + (cx - by^2)\,dx = 0,$$

$$by^2 - cx - \frac{ac}{2b} = Ce^{\frac{2bx}{a}}.$$

(94.) Let us now consider homogeneous differential equations. It may be proved that equations of this kind may always be rendered integrable by means of a homogeneous factor, and as the method of shewing this is just as easy for any number of variables as for two, we may as well take the more general case.

Let then the proposed equation be

$$du = Mdx + Ndy + Pdz + \&c. = 0 \,\,.\,.\,.\,.\,.\, (1),$$

which we shall consider to be homogeneous and inexact, and let U represent the factor which ought to render du an exact differential $d'u$; we shall then have

$$U\,du = U\,Mdx + U\,Ndy + U\,Pdz + \&c. = du' = 0 \,\,.\,.\,.\,.\, (2).$$

But from the property of homogeneous differential equations, demonstrated at (80), we have, by putting n for the degree of homogeneity of u',

$$U\,Mx + U\,Ny + U\,Pz + \&c. = nu' \,\,.\,.\,.\,.\, (3);$$

hence, dividing equation (1) by this, we have

$$\frac{Mdx + Ndy + Pdz + \&c.}{Mx + Ny + Pz + \&c.} = \frac{du'}{nu'} \,\,.\,.\,.\,.\, (4).$$

Now the second member of this equation is an exact differential, its integral being $\frac{1}{n}$ log. u', consequently this equation shews that the factor U, requisite to render (1) an exact differential, is

$$U = \frac{1}{Mx + Ny + Pz + \&c.}.$$

If there are but two variables, the requisite factor to render

$$Mdx + Ndy = 0$$

an exact differential is

$$\frac{1}{Mx + Ny}.$$

If the degree of homogeneity, n, of the sought integral should be 0, this process becomes inapplicable (80).

When it so happens that the factor thus deduced is ∞ or the denominator of U is 0, it becomes useless, as we require a finite factor, and this may in such cases be often otherwise discovered. Thus, taking the equation of two variables

$$Mdx + Ndy = 0,$$

if we find that

$$Mx + Ny = 0 \therefore N = - M \frac{x}{y},$$

so that the proposed may be put under the form

$$M \left\{ \frac{ydx - xdy}{y} \right\} = 0 \text{ or } M y\, d\, \frac{x}{y} = 0,$$

which will obviously be an exact differential if we multiply it by a factor U capable of rendering UMy equal to a function of $\frac{x}{y}$, and such a factor may often be readily discovered.

EXAMPLES.

1. To determine the integral of the differential equation

$$(yx + y^2)\, dx - (x^2 - yx)\, dy = 0.$$

In this example

$$M = yx + y^2, N = - x^2 + xy$$

$$\therefore \ \mathrm{U} = \frac{1}{\mathrm{M}x + \mathrm{N}y} = \frac{1}{2y^2 \, x} \, ;$$

therefore, multiplying the proposed by this factor, we have

$$\frac{dx}{2y} - \frac{xdy}{2\,y^2} + \frac{dx}{2x} + \frac{dy}{2y} = 0,$$

of which the integral is (79)

$$\frac{x}{y} + \log. \, xy = \mathrm{C}.$$

2. To determine the integral of the differential equation

$$(x^2 y + y^3) \, dx - (x^3 + xy^2) \, dy = 0.$$

Here

$$\mathrm{M} = y \, (x^2 + y^2), \ \mathrm{N} = - x \, (x^2 + y^2),$$

so that

$$\mathrm{M}x + \mathrm{N}y = 0,$$

and therefore, as above, the proposed may be put under the form

$$\mathrm{M} y \, d \, \frac{x}{y} = 0 = y \, (x^2 + y^2) \, d \, \frac{x}{y},$$

and we have now to discover what factor U, will make

$$\mathrm{U} \, y \, (x^2 + y^2) = \mathrm{F} \, \frac{x}{y}.$$

It is easy to perceive that $y^2 \, x^2$ is such a factor; multiplying by it, therefore, and we shall have, to integrate the equation,

$$\left(1 + \frac{y^2}{x^2}\right) d \, \frac{x}{y} = 0$$

or

$$\left(+ \frac{1}{z^2}\right) dz = 0.$$

The integral of this is $z - \dfrac{1}{z} = \mathrm{C}$, and, therefore, that of the proposed is

$$\frac{x^2 - y^2}{xy} = \mathrm{C}.$$

We need not multiply examples here, as the student may apply this process to the homogeneous equations, integrated at (88), by the separation of the variables. It is easy to shew that in every case in which the separation of the variables is possible, a factor may be found that will render the equation an exact differential, but we shall not seek this factor, as the process would comprehend only that class of differential equations which we know may be integrated by separating the variables.

As every differential equation of two variables is capable of being rendered an exact differential by means of a factor, and as, unfortunately analysis in its present state furnishes us with methods of finding this factor in but few cases, analysts, and especially *Euler*, have been induced to consider the inverse problem; that is, instead of seeking for the factor which would render a differential equation integrable, they have sought the relation which ought to exist among the variables and differentials of an equation, given in form only, in order that a factor given also in form might render it integrable. But, observes *Mr. Peacock,*[*] "these investigations frequently involve differential equations, which cannot be integrated by any known method, and it cannot be said that the cases in which they are successful are of very great importance or extent. In order to ensure this method all the success of which it is capable, it would require a very complete classification of the forms of differential equations of the first order, as well as a knowledge of the forms of the multipliers which are suited to each class. The immense extent, however, of this inquiry, and the difficulties which are met with, even in the simplest cases, preclude all hopes of its proving of much service in the general integration of differential equations of the first order." On the *inverse method of factors* the student may, however, consult *Dubourguet Calcul, Diff. et Int.* tom. 2, p. 88, and *Jephson's Calculus, vol.* 2, p. 145.

(95.) Before terminating the present chapter, we shall remark that we have sometimes to *differentiate under the sign of integration*, that is, M being a function of x and y, to determine $\dfrac{du}{dy}$ from $u = \int M dx$; this is done as follows:

---

[*] *Examples of the application of the Differential and Integral Calculus,* by George Peacock, A.M., FRS., &c. &c, p. 340.

Since

$$\frac{du}{dx} = \mathrm{M} \quad \text{and} \quad \frac{d^2u}{dy\,dx} = \frac{d^2u}{dx\,dy} = \frac{d\left(\frac{du}{dy}\right)}{dx} = \frac{d\mathrm{M}}{dy},$$

we have, by multiplying by dx, and integrating as regards x,

$$\frac{du}{dy} = \int \frac{d\mathrm{M}}{dy}\,dx.$$

CHAPTER IV.

ON THE GEOMETRICAL APPLICATIONS OF DIFFERENTIAL EQUATIONS OF THE FIRST ORDER AND DEGREE, AND ON EQUATIONS OF THE FIRST ORDER AND HIGHER DEGREES.

(96.) Before we proceed to consider differential equations of the higher orders and degrees, it will be desirable to present to the student a few geometrical problems of which the solutions depend upon the principles taught in the preceding chapter.

PROBLEM I.

To determine the curve whose tangent is a mean proportional between the part of the axis intercepted between it and a given point, and that same part augmented by a given line.

Let the rectangular axes originate at the given point, and let a denote the given line. The distance of the origin from the intersection of the axis with the tangent will be expressed by the difference between the subtangent and abscissa, that is, by

$$y\frac{dx}{dy} - x;$$

hence, by the conditions of the problem, the differential equation of the required curve is

$$(y \frac{dx}{dy})^2 + y^2 = (y \frac{dx}{dy} - x)(y \frac{dx}{dy} - x + a)$$

or

$$y^2 = (a - 2x) y \frac{dx}{dy} - (a - x) x$$

or

$$dy = \frac{(a - 2x) y dx - (a - x) x dy}{y^2},$$

and, as the second member is an exact differential, we have, by integrating,

$$y = \frac{(a - x) x}{y} + C \therefore y^2 = (a - x) x + Cy,$$

therefore the equation of the curve is

$$y^2 + x^2 - (ax + Cy) = 0,$$

which is, therefore, a circle passing through the origin, and of which the coordinates of the centre are $\frac{1}{2} a$, $\frac{1}{2} C$, (*Anal. Geom.* p. 39.) The determination of C requires an additional condition.

PROBLEM II.

To determine the curve of which the normal is equal to that part of the axis of x intercepted between it and the origin.

The part of the axis between the normal and the origin is the sum of the subnormal and abscissa, that is, it is

$$y \frac{dy}{dx} + x,$$

also the expression for the normal is

$$\sqrt{y^2 + y^2 \frac{dy^2}{dx^2}} ;$$

hence the differential equation of the curve is

$$y^2 + y^2 \frac{dy^2}{dx^2} = (y \frac{dy}{dx} + x)^2$$

or

$$y^2 - 2xy\,\frac{dy}{dx} - x^2,$$

or

$$y^2\,dx - 2xy\,dy - x^2\,dx.$$

This equation is homogeneous but not an exact differential; hence, by (94), the integral is

$$\frac{(y^2 + x^2)}{2x} = C \therefore y^2 + x^2 - 2Cx = 0,$$

consequently the required curve is a circle passing through the origin, and of which the radius is C, any arbitrary line.

On Trajectories.

(96.) A trajectory is a curve which intersects a given family of curves all in the same constant angle.

Let the general equation of any family of curves be

$$F'(x, y, a) = 0 \ .\ .\ .\ .\ (1),$$

a being the arbitrary parameter, and let the sought curve be such as to intersect each of these in the constant angle tan.$^{-1}$ x. Let us first consider some individual curve of the family (1), a having a fixed value, then putting p' for the $\dfrac{dy}{dx}$ derived from its equation, in order to distinguish it from the $\dfrac{dy}{dx}$ derived from the equation of the sought curve, we have (*Anal. Geom.* p. 25)

$$a = \frac{\dfrac{dy}{dx} - p'}{1 + p'\dfrac{dy}{dx}} \ .\ .\ .\ .\ (2).$$

$$\therefore \frac{dy}{dx} = \frac{p' + a}{1 - ap'} \ .\ .\ .\ .\ (3).$$

Now whatever be the sought curve, this equation, in conjunction with (1) will obviously determine the point (x, y) of intersection with the individual curve.   It follows, therefore, that if we eliminate the parameter a, by means of these equations, the result will be the locus of these points for all the curves of the family (1), that is to say, it will, be the equation of the trajectory sought.

If the constant angle of intersection is a right angle, then $a = \infty$, and consequently from (2)

$$1 + p'\frac{dy}{dx} = 0 \ldots \ldots (3'),$$

and eliminating a, by means of this and (1), the resulting equation will be that of the rectangular trajectory.

It may be here remarked that instead of leaving the elimination of a till we come to the equation (3) or (3'), we may previously perform the elimination by means of (1) and its differential p', since p' is the only term in (3) into which it can enter.

PROBLEM III.

To determine the curve which intersects at a right angle every straight line of the family

$$y = ax \ldots \ldots (1),$$

that is to say, all the straight lines that can be possibly drawn through the origin.

By differentiating this equation, we have

$$p' = a \ldots \ldots (2),$$

therefore, eliminating a by means of this and (1), we have

$$p' = \frac{y}{x} \therefore (3'),\ 1 + p'\frac{dy}{dx} = 1 + \frac{y}{x} \cdot \frac{dy}{dx} = 0$$

$$\therefore x\,dx + y\,dy = 0,$$

which is the differential equation of the trajectory; hence, by integrating,

$$x^2 + y^2 = C,$$

so that the trajectory is a circle of arbitrary radius.

PROBLEM IV.

To determine the trajectory which intersects the series of straight lines

$$y = ax$$

in the oblique angle $\tan.^{-1}a$.

As before,

$$p' = \frac{y}{x},$$

hence the equation (3) is

$$\frac{dy}{dx} = \frac{\frac{y}{x} + a}{1 - a\,\frac{y}{x}} \quad \therefore (x - ay)\,dy - (y + ax)\,dx = 0,$$

which is the differential equation of the sought curve. By integrating this (89), we have

$$\tan.^{-1}\frac{y}{x} = a \log. C \sqrt{x^2 + y^2},$$

but a times the hyp. log. of any quantity is equal to the log. of the same quantity in the system whose modulus is a; hence, calling the base corresponding to this modulus b, the foregoing equation is the same as

$$\text{Log. } b \tan.^{-1}\frac{y}{x} = \text{Log. } C \sqrt{x^2 + y^2},$$

or, putting

$$\sqrt{x^2 - y^2} = r \text{ and } \tan.^{-1}\frac{y}{x} = \omega$$

$$b^{\omega} = Cr,$$

C being arbitrary. If we assume it so that $r = 1$ when $\omega = 0$, then $C = 1$, and the equation becomes

$$b^{\omega} = r,$$

which is that of a logarithmic spiral; b being the base of the system represented, $a = \tan. \angle P$ *(see Diff. Calc.* p. 118,*)* the modulus, and $\tan.^{-1}\frac{y}{x} = \omega$ being the angle PFA.

PROBLEM V.

To determine the rectangular trajectory of the system of parabolas

$$y^2 = 2ax.$$

Here

$$p' = \frac{a}{y} \text{ and } a = \frac{y^2}{2x} \therefore p' = \frac{y}{2x}.$$

Hence the equation (3') is

$$1 + \frac{y}{2x} \cdot \frac{dy}{dx} = 0 \therefore 2x\,dx + y\,dy = 0,$$

which is the differential equation. Integrating this, we have

$$2x^2 + y^2 = C,$$

the equation of an ellipse, of which the centre is at the common vertex of the variable parabolas, and of which the axes are $2\sqrt{C}$ and $\sqrt{2C}$. As C is arbitrary there are, as usual, an infinite number of elliptic trajectories, but in all the axes are to each other as 2 to $\sqrt{2}$, or as $\sqrt{2}$, to 1.

PROBLEM VI.

To determine the rectangular trajectory of the series of parabolas whose general equation is

$$y^n = ax^m.$$

The trajectory is the ellipse $my^2 + nx^2 = C.$

PROBLEM VII.

To determine the rectangular tajectory of a series of circles all touching a given straight line at a given point.

Any circle passing through the given point, and having its centre on the given line is a rectangular trajectory.

Integration of Differential Equations of the First Order and of the Higher Degrees.

(97.) The most general form of a differential equation containing two variables, and of the nth degree, is

$$\frac{dy^n}{dx^n} + \mathrm{P}\,\frac{dy^{n-1}}{dx^{n-1}} + \ldots + \mathrm{M}\,\frac{dy}{dx} + \mathrm{N} = 0 \ldots (1),$$

and such an equation we know (82) cannot be the immediate differential of any integral, but must be derived from its primitive by the elimination of a constant which enters it in the nth degree. This elimination may be considered to be performed thus. The primitive being solved for the proposed constant C we shall have, in consequence of C having n roots, n expressions for C, in terms of x and y, from all of which C will vanish by differentiation, and we shall thus have n differential equations of the first degree of which the integral of each will satisfy, being indeed a factor of, the primitive, and their product will be the equation of the nth degree (1).

Hence, to return from (1) to the primitive, we must find its n component factors of the first degree in $\dfrac{dy}{dx}$, integrate each of these annexing the same constant C and then multiply the n results together, and we shall thus have the complete primitive. The theory of this class of equations is therefore very easy and obvious, but the resolution of (1) into its component simple factors, or, in other words, the solution of an equation of the nth degree, is a problem not to be accomplished in the present state of analysis, except in a few particular cases. We shall give an example or two in these cases.

EXAMPLES.

(98.) 1. Given

$$y\,\frac{dy^2}{dx^2} + 2x\frac{dy}{dx} - y = 0.$$

This being an equation of the second degree, the two values of $\dfrac{dx}{dy}$

are determinable: they are

$$\frac{dy}{dx} = \frac{-x \pm \sqrt{y^2 + x^2}}{y},$$

hence the component factors of the proposed are

$$y \frac{dy}{dx} + x + \sqrt{y^2 + x^2} = 0$$

$$y \frac{dy}{dx} + x - \sqrt{y^2 + x^2} = 0$$

which reduce to

$$\pm \frac{y\, dy + x\, dx}{\sqrt{y^2 + x^2}} = dx,$$

and it is plain that the first member of this is the differential of $\pm \sqrt{y^2 + x^2}$; consequently the factors of the required primitive are

$$\pm \sqrt{y^2 + x^2} = x + C,$$

of which the product is

$$y^2 = 2Cx + C^2.$$

It is easy to see, without further illustration, how the primitive is to be obtained, when the proposed differential equation is resolvable into its constituent factors. When this resolution is impossible, the primitive may nevertheless be obtained by analytical artifice, when the proposed appears under certain forms.

I.

The solution may be effected when only one of the variables x or y enter the proposed, provided the equation can be solved for this variable.

Put p' for the differential coefficient, then, as the equation contains but one of the variables, say x, and as moreover it may be solved for this, we may reduce it to the form

$$x = Fp' \quad \ldots \ldots (1).$$

Now since $dy = p'dx$, we have, by integrating by parts, the second member,

$$y = xp' - \int x dp' \ldots \ldots (2).$$

Substituting (1) in (2), we have the equation

$$y = p' \, Fp' - \int Fp' \, dp' \ldots \ldots (3),$$

it remains, therefore, to integrate the differential of a single variable $Fp'dp'$, and then to eliminate p' by means of (1) and (3); the result will be the sought relation between x and y.

It may be here remarked that if the proposed equation is not so easily solvable for x, as for p', then, instead of (1), we had better get

$$p' = Fx \therefore dy = Fx \cdot dx,$$

which immediately gives the required relation

$$y = \int Fx \cdot dx.$$

2. Given the equation

$$x \, \frac{dy^2}{dx^2} + x - 1 = 0, \text{ or } xp'^2 + x - 1 = 0$$

to determine the relation between x and y.

Solving for x, we have

$$x = \frac{1}{p'^2 + 1} \ldots \ldots (1.)$$

Hence the equation (3) is

$$y = \frac{p'}{p'^2 + 1} - \int \frac{dp'}{p'^2 + 1}$$

$$= \frac{p'}{p'^2 + 1} - \tan.^{-1} p' + C \ldots \ldots (2),$$

putting in this the value of p', furnished by (1), we have

$$y = \sqrt{x - x^2} - \tan.^{-1} \sqrt{\frac{1 - x}{x}} + C.$$

If, in the case we are now considering, the proposed is not solvable, either for x or p', then the artifice usually employed is that of substituting xz for p', as by this means we obtain an equation of which all the terms, unless one is constant, become divisible by a power of x, and therefore the degree of the equation may be depressed. If then, in this depressed state, the equation can be solved for x in terms of z, or for z in terms of x, we shall have, by substituting the result in the assumed condition $p' = xz$, either

$$\frac{dy}{dfz} = zfz \therefore y = \int z\,fz\,dfz \ \ldots \ldots \ (1),$$

or else

$$\frac{dy}{dx} = xfx \therefore y = \int x\,fx\,dx \ \ldots \ldots \ (2).$$

In the first case the relation between x and y will be given by combining (1) with the depressed equation.

3. Given the equation

$$y^5 \frac{dy^5}{dx^5} + y^2 \cdot \frac{dy^3}{dx^3} + 1 = 0.$$

Here, if we were to substitute yz for $\dfrac{dy}{dx}$, we should be no more able to depress the equation than in its present form, because of the constant 1; but if we change its form by multiplying the terms by $\dfrac{dx^5}{dy^5}$, it becomes

$$y^5 + y^2 \frac{dx^2}{dy^2} + \frac{dx^5}{dy^5} = 0,$$

which, by the substitution of yz for $\dfrac{dx}{dy}$, reduces to

$$y^5 + y^4 z^2 + y^5 z^5 = 0 \text{ or } y + z^2 + yz^5 = 0$$

$$\therefore y = - \frac{z^2}{1 + z^5} \ \ldots \ldots \ (1)$$

$$\therefore dy = - \frac{z\,(2 - 3z^5)\,dz}{(1 + z^5)^2};$$

hence, in virtue of the condition

$$dx = yz\,dy,$$

we have, by substitution,

$$dx = \frac{2z^4 dz}{(1 + z^5)^3} - \frac{3z^9 dz}{(1 + z^5)},$$

and, integrating,

$$x = \frac{3}{5\,(1+z^5)} - \frac{1}{2\,(1+z^5)^2} + C \ldots \ldots (2).$$

The equations (1) and (2), combined, express the relation between x and y. To eliminate z we may first determine $1 + z^5$ from the quadratic (2), and thus obtain the function of x, which equals the denominator of (1); and if 1 be taken from this function, the $\frac{2}{3}$ power of the result will be the numerator.

II.

The solution may be effected when both variables enter the proposed, provided they render the terms homogeneous with respect to the variables, and provided, moreover, we could solve the equation for x, if it were equal to y.

For, let n be the degree of homogeneity, then, by substituting xz for y, and dividing by x^n, which must necessarily be a common factor of the terms, we shall have an equation between z and p', in which the highest power of z will be n; if then this equation can be solved for z, we shall have

$$z = Fp' \,\therefore\, dz = dFp',$$

but, since

$$y = xz \,\therefore\, dy = xdz + zdx.$$

or, substituting for z and dz the values above,

$$dy = xdFp' + Fp' \cdot dx,$$

or, since $dy = p'dx$, this equation reduces to

$$(p' - Fp')\, dx = xdFp' \,\therefore\, \frac{dx}{x} = \frac{dFp'}{p' - Fp'},$$

whence

$$\log . \, x = \int \frac{dFp'}{p' - Fp'},$$

and this, in conjunction with the assumed condition

$$y = xFp',$$

furnishes the required relation between x and y.

4. Given the equation

$$y - xp' = x\sqrt{1 + p'}$$

to determine the relation between x and y.

Putting xz for y, and dividing the result by x, we have

$$z - p' = \sqrt{1 + p'^2} \therefore z = p' + \sqrt{1 + p'^2}$$

$$\therefore dz = dp' + \frac{p'dp'}{\sqrt{1 + p'^2}},$$

and from the equation

$$p'dx = dy = xdz + zdx$$

we get

$$\frac{dx}{x} = \frac{dz}{p' - z} = -\frac{dp'}{\sqrt{1 + p'^2}} - \frac{p'dp'}{1 + p'^2},$$

and, integrating

$$\log. x = -\log. (p' + \sqrt{1 + p'^2}) - \log. \sqrt{1 + p'^2} + \log. C;$$

hence

$$x = \frac{C}{\sqrt{1 + p'^2}\,(p' + \sqrt{1 + p'^2})},$$

and this, in conjunction with the assumed condition

$$y = xz = x\,(p' + \sqrt{1 + p'^2})$$

expresses the relation between x and y. To get this in a single equation, we must eliminate p'; and, in order to this, substitute the value of x, above, in this expression for y, and it becomes

$$y = \frac{C}{\sqrt{1 + p'^2}} \therefore 1 + p'^2 = \frac{C^2}{y^2} \therefore p' = \sqrt{\frac{C^2}{y^2} - 1},$$

hence, by substitution, the expression above for x becomes

$$x = \frac{y^2}{C + \sqrt{C^2 - y^2}},$$

the relation required.

III.

Another integrable form is

*airaut's
orm.*

$$y = x\frac{dy}{dx} + \mathrm{F}\,\frac{dy}{dx} \;.\;.\;.\;.\;(1),$$

in which the function $\mathrm{F}\dfrac{dy}{dx}$ contains neither x nor y. This is *Clairaut's
form.*

By differentiating this form, we have

$$\frac{dy}{dx} = p' = p' + x\,\frac{dp'}{dx} + \frac{d\mathrm{F}p'}{dp'}\,\frac{dp'}{dx}$$

$$\therefore 0 = \left\{ x + \frac{d\mathrm{F}p'}{dp'} \right\}\frac{dp'}{dx'} \;.\;.\;.\;.\;(3),$$

which equation leads equally to the two conditions

$$(4)\;.\;.\;.\;.\; x + \frac{d\mathrm{F}p'}{dp'} = 0 \;\text{and}\; \frac{dp'}{dx} = 0 \;.\;.\;.\;.\;(5).$$

Now the first of these contains no differential, for $\dfrac{d\mathrm{F}p'}{dp'}$ is the differential *coefficient* of $\underline{dp'}$, and is, therefore, a function of p', so that, if p'
be eliminated by means of (1) and (2), the resulting equation between
x and y will certainly satisfy the proposed, but yet cannot be the complete primitive, since no arbitrary constant is introduced. The complete primitive must, therefore, be furnished by the other condition (5).

Now the condition $\dfrac{dp'}{dx} = 0$ leads to $p' = \mathrm{C}$; hence the remarkable
fact, that in Clairaut's form the complete primitive is found, by merely
substituting the arbitrary constant C for p' in that form.

EXAMPLE.

To determine the complete primitive of the equation

$$y - x\frac{dy}{dx} = a + a\frac{dy^2}{dx^2},$$

or

$$y = p'x + a(1 + p'^2).$$

Substituting C for p', we find for the primitive the equation

$$y = Cx + a\,(1 + C^2),$$

as is obvious, for if we differentiate this, we get $p' = C$, and this put for C, in the equation, produces the proposed.

In like manner, the complete primitive of

$$y\,dx - x\,dy = a\,(dx^3 + dy^3)^{\frac{1}{3}},$$

or

$$y = p'x + a\,(1 + p'^3)^{\frac{1}{3}},$$

is

$$y + Cx + a\,(1 + C^3)^{\frac{1}{3}}.$$

IV.

We shall terminate the present chapter by exhibiting the integral of the form

$$y = Px + Q,$$

where P and Q are functions of p'.

By differentiating, we have

$$dy = p'dx = Pdx + xdP + dQ$$

$$\therefore (P - p')\,dx + xdP + dQ = 0$$

$$\therefore dx + x\,\frac{dP}{P - p'} = -\,\frac{dQ}{P - p'}.$$

This last is a linear equation (87), and, therefore,

$$x = e^{-\int \frac{dP}{P-p'}} \left\{ \int e^{\int \frac{dP}{P-p'}}\,\frac{dQ}{P-p'} \right\},$$

consequently this equation, in conjunction with the proposed, expresses the relation between x and y, and if p' be eliminated therefrom, this relation will be expressed in a single equation between the variables.

CHAPTER V.

ON THE THEORY OF SINGULAR SOLUTIONS OF DIFFE-RENTIAL EQUATIONS OF THE FIRST ORDER.

(99.) Every differential equation may be considered to have been derived from its primitive, by eliminating a constant between it and its immediate differential. Thus, if F $(x, y, c) = 0$, c being the arbitrary constant, is the complete primitive of the differential equation

$$f\left(x, y, \frac{dy}{dx}\right) = 0 \ . \ . \ . \ . \ (1),$$

then has (1) arisen from eliminating c, by means of the equations

$$(2) \ . \ . \ . \ . \ F\left(x, y, c\right) = 0, \ \frac{dF\left(x \ y, c\right)}{dx} = 0 \ . \ . \ . \ . \ (3).$$

Now if instead of the constant quantity c, any variable quantity were to be substituted in each of these equations the result of the elimination of that variable would obviously be the same equation (1), so that, if c be supposed to be such a variable that the differential (3) of (1) may be precisely the same as when we supposed it constant, then this variable value may be attributed to c, without affecting in anywise the result (1) of its elimination. Let us then see whether it is possible for such a variable value of c to exist. By considering c variable, as well as x and y, the differential coefficient derived from (2), relatively to the independent variable x, is

$$\frac{dF\left(x, y, c\right)}{dx} + \frac{dF\left(x, y, c\right)}{dc} \ \frac{dc}{dx} = 0,$$

and in order that this expression may be the same as would arise from differentiating, on the supposition of c constant, that is, in order that it may be identical to (3), it is obviously merely necessary to determine c from the condition

$$\frac{dF\left(x, y, c\right)}{dc} \ \frac{dc}{dx} = 0,$$

which condition may be satisfied upon either of the hypotheses

$$(4) \ldots \ldots \frac{dc}{dx} = 0 \text{ or } \frac{d\mathrm{F}\,(x, y, c)}{dc} = 0 \ldots \ldots (5).$$

The first fixes a *constant* value for c, and therefore the requisite *variable* value is to be determined from the second. This variable then put for c, in (2), however it may alter the form or degree of that equation does not deprive it of the character of being an integral of (1) seeing that this last arises from the combination of (2) and its immediate differential (3). This integral is necessarily different from the complete primitive (2), since in this c is an arbitrary constant, while in the other case, it is a certain function of x and y, determinable from (5).

We see, therefore, that it is possible for a differential equation to have other integrals besides the complete primitive, but derivable from it by substituting in it, for the arbitrary constant c, each of its values given in terms of x and y by the equation (5). Such integrals are called *singular integrals*, or *singular solutions* of the proposed differential equation.

It must be here particularly remarked, that the value of c, as deduced from the equation (5), is not necessarily a function of the variables; for c may be connected with these variables in $\mathrm{F}(x, y, c)$ merely by way of addition or subtraction, in which case (5) will imply $fc = 0$, the roots of which equation will be particular constant values of c, which, substituted in the complete primitive, will furnish so many *particular* cases of that primitive; these, therefore, will be but *particular solutions*. Moreover the value of c, as deduced from (5), may appear under the form of a function of x and y, and yet be, in reality, a constant value; for the complete primitive, if solved for one of the constants a, which enter it, will furnish for the value of that constant a function of $x, y,$ and c; if, therefore, this function, by assuming any particular value for c, or, indeed, if any function ϕ of this function, agree with the function for c given by (5), then the substitution of this latter for c in the primitive is no more than substituting the constant ϕa, and thus the solution is not a *singular*, but, as before, a *particular* solution, and would have been immediately furnished by the primitive, upon substituting ϕa for c. It is necessary, therefore, before we pronounce the result of the elimination of c from the equations (2) and (5) to be a singular solution of (1), to assure ourselves that this same result cannot be obtained by the mere substitution of a constant function for c in (2).

It may be here remarked, that if the value of y or of x be deduced from the complete primitive 2, we may write it

$$y + f(x, c) = 0 \text{ or } x + f(y, c) = 0;$$

hence the differential of either of these with respect to c only, that is, the condition (5) becomes either

$$\frac{df(x, c)}{dc} = 0, \text{ or } \frac{df(y, c)}{dc} = 0,$$

that is,

$$\frac{dy}{dc} = 0, \text{ or } \frac{dx}{dc} = 0 \ldots \ldots (6),$$

so that the values of c, corresponding to singular solutions, are furnished equally by equation (5), or by these two.

There is one class of differential equations which we can at once affirm to have no singular solution, viz. those into whose complete primitives the arbitrary constant c enters only in the first power; for in such cases c will be eliminated in (5) by differentiation, so that this equation fails in this case to supply a value for c. We have seen (82) that equations of the first order and nth degree arise from primitives into which c enters in the nth degreee. Hence no differential equation of the first order and degree can have a singular solution.* *See Note* D.

Before proceeding further, let us illustrate what has been said by an example, and let the proposed equation be

$$ydx - xdy = a \sqrt{dx^2 + dy^2},$$

or

$$y = p'x + a \sqrt{1 + p'^2},$$

which being of Clairaut's form, its complete primitive is

$$y = Cx + a \sqrt{1 + C^2} \ldots \ldots (1).$$

---

* It must not be forgotten, that in all our reasonings on the theory of differential equations, they are considered as freed from radicals.

To determine the singular solution, we are to eliminate C between this result and

$$\frac{dy}{dc} = x + \frac{ac}{\sqrt{1 + c^2}} = 0 \therefore c = -\frac{x}{\sqrt{a^2 - x^2}},$$

substituting this in (1), we have

$$y = \sqrt{a^2 - x^2},$$

or

$$y^2 + x^2 = a^2 \ldots \ldots (2),$$

which is the singular solution, for it can never be comprised in the complete primitive (1), since whatever value we give to c, that equation always represents a straight line, while (2) represents a circle.

As in Clairaut's form, to which the above example belongs, the complete primitive is always the same as the proposed differential equation, viewing the coefficient p' in the light of an arbitrary constant, it is evident that in this form the singular solutions may be obtained by eliminating p' between the proposed and $\dfrac{dy}{dp'} = 0$, or $\dfrac{dx}{dp'} = 0$. If this were the case with other forms as well as with that of Clairaut, we should then be able to determine the singular solutions whenever they exist from the proposed differential equation, without being at the trouble of first finding the complete primitive. Let us then examine this point.

(100.) It has been seen that the differential equation (3) is the same, whether c be constant causing (2) to be the complete integral of (1), or whether it be such a variable as to cause (2) to be the singular solution of (1). In either case the elimination of c from these two equations produces (1), so that if we solve (3) for c, calling the result

$$c = \phi(x, y, p'),$$

and substitute this value in (2) we shall have (1) under the form

$$u = F(x, y, \phi) \ldots \ldots (1'),$$

where ϕ is put for $\phi(x, y, p')$.

This equation being the same as (1), the original differential equation, it follows that if we substitute for p' which enters the function ϕ, its value as deduced from (1), when put under the form

$$p' + f(x, y) = 0,$$

that is to say, the value

$$p' = -f(x, y),$$

the expression for u, (1'), will be identically 0, that is, independently of any relation between x and y. As, therefore, (1') fixes no relation between x and y, we may differentiate this equation as if they were independent, taking care to observe that ϕ is a function of x, y and p', and that $p' = -f(x, y)$; hence, differentiating with respect to x, we have

$$\frac{du}{dx} + \frac{du}{d\phi} \frac{d\phi}{dx} - \frac{du}{d\phi} \frac{d\phi}{dp'} \frac{dp'}{dx} = 0,$$

and with respect to y,

$$\frac{du}{dy} + \frac{du}{d\phi} \frac{d\phi}{dy} - \frac{du}{d\phi} \frac{d\phi}{dp'} \frac{dp'}{dy} = 0,$$

From these two equations we get

$$\frac{dp'}{dx} = \left(\frac{du}{dx} + \frac{du}{d\phi} \frac{d\phi}{dx} \right) \div \frac{du}{d\phi} \frac{d\phi}{dp'},$$

$$\frac{dp'}{dy} = \left(\frac{du}{dy} + \frac{du}{d\phi} \frac{d\phi}{dy} \right) \div \frac{du}{d\phi} \frac{d\phi}{dp'}.$$

Now, in the case of a singular solution, we must have

$$\frac{du}{d\phi} = 0,$$

for then the value ϕ of c is determined conformably to the condition

$$\frac{du}{dc} = 0,$$

and consequently the two foregoing equations become

$$\frac{dp'}{dx} = \infty, \quad \frac{dp'}{dy} = \infty,$$

$$\therefore \frac{dx}{dp'} = 0, \quad \frac{dy}{dp'} = 0,$$

hence, if p' be eliminated by means of either of these and the proposed differential equation, the result will be a singular solution, if it be a solution at all, that is, if it satisfy the proposed equation.

The preceding conditions lead to another for the determination of p', sometimes of more convenient application than those. Thus the proposed differential equation being

$$U = f(x, y, p') = 0,$$

we have, by differentiating it,

$$\left\{\frac{dU}{dx}\right\} = \frac{dU}{dx} + \frac{dU}{dy}\,\frac{dy}{dx} + \frac{dU}{dp'}\left(\frac{dp'}{dx} + \frac{dp'}{dy}\,\frac{dy}{dx}\right) = 0$$

$$\therefore \frac{dU}{dp'} = -\left(\frac{dU}{dx} + \frac{dU}{dy}\,\frac{dy}{dx}\right) \div \left(\frac{dp'}{dx} + \frac{dp'}{dy}\,\frac{dy}{dx}\right)$$

But by the foregoing conditions, this divisor is ∞ ; hence

$$\frac{dU}{dp'} = 0,$$

which equation will give the values of p', necessary to fulfil the conditions above. It must be remarked that throughout this article u has been considered as a function of both x and y, x being the independent variable; but the singular solution $u = 0$ may contain only x, which cannot, of course, satisfy the proposed, but by considering y as the independent variable; hence for such solutions as these, we must, in the foregoing condition, consider $p' = \dfrac{dx}{dy}$.

(101.) The connexion between the complete primitive and the singular solution is susceptible of geometrical illustration. For the complete primitive represents always a family of curves, c being the variable parameter, and we know (*Diff. Calc.* p. 145,) that the envelope of this family is analytically represented by the equation which arises from eliminating c by means of the complete primitive, and its differential with respect to c. But we have seen that the singular solution is given by the same elimination: hence the singular solution is the equation of the curve which envelopes the family represented by the complete primitive. As in Clairaut's form, the complete primitive is the equation of a family of straight lines, it follows that if this form ought to belong to a curve, the equation of that curve must be the singular solution.

(112.) We shall now add a few examples of the determination of singular solutions.

EXAMPLES.

1. Given the equation

$$U = (x + y) p' - x p'^2 - (a + y) = 0$$

to determine the singular solution

$$\frac{dU}{dp'} = x + y - 2xp' = 0$$

$$\therefore p' = \frac{x + y}{2x}.$$

Substituting this in $U = 0$, it becomes

$$\frac{(x + y)^2}{2x} - \frac{(x + y)^2}{4x} - (a + y) = 0$$

or

$$x - y - 2\sqrt{ax} = 0.$$

If this satisfies the proposed, it is the singular solution.   In order to ascertain this, substitute in the proposed

$$y = x - 2\sqrt{a - x}$$

$$p' = 1 - \frac{a}{\sqrt{ax}},$$

and we find the first member become

$$2 (x - \sqrt{ax}) \left(1 - \frac{a}{\sqrt{ax}}\right) - \frac{(\sqrt{ax} - a^2)}{a} - (a + x - 2\sqrt{ax})$$

on

$$2 (x - \sqrt{ax}) - 2 (\sqrt{ax} - a) - x + 2\sqrt{ax} - a - a - x + 2\sqrt{ax},$$

where it is obvious that the terms destroy each other; hence the above is the singular solution.

2. Given the equation

$$y p'^2 + 2p'x - y = 0$$

to determine the general and the singular solution.

By solving the equation for p', we have found (p. 225) the complete integral to be

$$y^2 - 2cx - c^2 = 0,$$

and, differentiating with respect to c, we have

$$x + c = 0 \therefore c = -x,$$

and this, substituted in the primitive, furnishes the singular solution

$$y^2 + x^2 = 0.$$

3. Given the equation

$$U = x^2 + 2xyp' + (a^2 - x^2) p'^2 = 0$$

to determine the singular solution

$$\frac{dU}{dp'} = 2xy + 2 (a^2 - x^2) p' = 0 ;$$

to eliminate p' by means of these equations, multiply the latter by p' and subtract it from the former, and we have

$$x^2 + xy\, p' = 0 \therefore p' = -\frac{x}{y} ,$$

and this substituted in the proposed, gives

$$x^2 + y^2 - a^2 = 0,$$

an equation which satisfies the proposed, and which is therefore the singular solution.

4. Given the equation

$$U = (x^2 - 2y^2) p'^2 - 4xy\, p' - x^2 = 0$$

to determine the singular solution

$$\frac{dU}{dp'} = (x^2 - 2y^2) p' - 2xy = 0.$$

Eliminating p' by means of these equations there results

$$x^2 (x^2 + 2y^2) = 0,$$

which is satisfied by either

$$x^2 = 0 \text{ or } x^2 + 2y^2 = 0,$$

but only the latter satisfies the proposed equation: this, therefore, is the singular solution.

5. Given the equation

$$x\,dy - y\,dx = dx \sqrt{x^2 + y^2}$$

to determine the singular solution

$$x^2 = - y^2.$$

6. Given the equation

$$y\,dx - x\,dy = x \sqrt{dx^2 + dy^2}$$

to prove that there is no singular solution.

7. Given the general solution or complete primitive

$$y = x + (c - 1)^2 (c - x)^2$$

to prove that the only singular solution is that corresponding to $c = \frac{1}{2}(x + 1)$.

(103.) We shall conclude the present chapter with one or two geometrical problems which conduct to singular solutions.

PROBLEM I.

To find a curve such that the perpendiculars drawn from a given point upon its tangents may be all of the same constant length.

Let (x, y) represent in general any point in the required curve, then the equation of the tangent through it will be (*Diff. Calc.* p. 112,)

$$Y - y = \frac{dy}{dx}(X - x) \text{ or } Y = p'X + y - p'x,$$

and supposing the given point to be the origin, the perpendicular from it on this line will be expressed by (*Anal. Geom.* p. 29,)

$$a = \frac{y - p'x}{\sqrt{p'^2 + 1}},$$

which being constant we have

$$y = p'x + a\sqrt{p'^2 + 1}$$

for the differential equation of the required curve.

The complete integral of this equation is (p. 230,)

$$y = cx + a\sqrt{1 + c^2} \ .\ .\ .\ .\ (1),$$

which represents a family of straight lines, and the general expression for the perpendicular from the origin on any one of them is

$$\frac{y - cx}{\sqrt{c^2 + 1}}.$$

which is equal to a, the constant length.

Now the singular solution of the proposed is

$$y^2 + x^2 = a^2 \ .\ .\ .\ .\ (2),$$

which represents a circle, and since the radius is a it is plain that it touches all the straight lines whose perpendicular distance from the centre is a; hence, agreeably to (101), the singular solution (2) touches and envelopes all the particular solutions comprised in (1).

PROBLEM II.

To find a curve such that the product of the two perpendiculars drawn from two given points on any tangent may be constant.

Let the axis of x pass through the given points and take the origin at the middle point between them, so that the abscissas of the points will be a and $-a$. Then the expression for the perpendicular from the point $(a, 0)$ on the line

$$Y = p'X + y - p'x$$

is (*Anal. Geom.* p. 29,)

$$\frac{-p'a - (y - p'x)}{\sqrt{p^2 + 1}} = -\frac{y + p'(a - x)}{\sqrt{p^2 + 1}},$$

and from the point $(-a, 0)$ on the same line

$$- \frac{y - p'(a + x)}{\sqrt{1 + p^2}}.$$

The product of these two expressions is to be constant,

$$\therefore \frac{y^2 - p'^2(a^2 - x^2) - 2p'xy}{p^2 + 1} = b^2,$$

this equation solved for y gives

$$y = p'x \pm \sqrt{b^2 + m^2 p'^2},$$

m^2 being put for $a^2 + b^2$. This equation being of Clairaut's form, we have for the complete primitive

$$y = cx \pm \sqrt{b^2 + m^2 c^2},$$

which represents a system of straight lines. The singular solution, or the equation of the curve to which these are tangents, is

$$m^2 y^2 + b^2 x^2 = m^2 b^2.$$

The curve sought is therefore an ellipse.

PROBLEM III.

To find a curve such that the normal may have a constant ratio to the part of the axis intercepted between it and the origin.

The curve may be either a circle or a parabola.*

---

* For a more comprehensive view of the theory of Singular Solutions the student is referred to the *Calcul des Fonctions,* where *Lagrange* has devoted upwards of 100 pages to this subject.

CHAPTER VI.

ON THE INTEGRATION OF DIFFERENTIAL EQUATIONS OF THE SECOND AND HIGHER ORDERS.

(104.) The most general form of a differential equation of the second order is

$$F\left(x, y, \frac{dy}{dx}, \frac{d^2 y}{dx^2}\right) = 0 \ . \ . \ . \ . \ (A),$$

which, however, comprehends a great variety of cases that are not integrable by any general process. Under certain conditions the integration is always possible, or may at least be reduced to the integration of an inferior order: as for example when the function does not contain all four of the quantities x, y, $\dfrac{dy}{dx}$, $\dfrac{d^2 y}{dx^2}$; also when the equation is homogeneous with respect to the variables and the differentials, and in one or two other cases. It may be remarked here that in integrating equations of the higher orders we have not the option of making which we choose the independent variable, as in equations of the first order, without altogether altering the form of the equation, for by changing the independent variable the second differential coefficient will be supplied by a function of more complicated form, although such a change sometimes facilitates integration.

Let us now examine those cases of the general differential equation of the second order which are integrable by general process, and first those into which all four of the quantities within the parentheses do not enter. We shall thus have five classes of equations, viz. three containing but two of these quantities or of the forms

$$F\left(x, \frac{d^2 y}{dx^2}\right) = 0, \ F\left(y, \frac{d^2 y}{dx^2}\right), \ F\left(\frac{dy}{dx}, \frac{d^2 y}{dx^2}\right) = 0,$$

and two into which three of the quantities enter; their forms being

$$F\left(x, \frac{dy}{dx}, \frac{d^2 y}{dx^2}\right) = 0, \ F\left(y, \frac{dy}{dx}, \frac{d^2 y}{dx^2}\right).$$

I.

To integrate the form

$$F\left(x, \frac{d^2 y}{dx^2}\right) = 0.$$

Solve this equation for $\frac{d^2 y}{dx^2}$ and we shall have

$$\frac{d^2 y}{dx^2} = fx \ \therefore \ d^2 y = fx\,dx^2 = X\,dx^2$$

$$\therefore \ y = \int^2 X\,dx^2 = X_{,} + C_{,}\,x + C,$$

where $X_{,}$ represents the second integral of $X\,dx^2$ without the arbitrary constants, (see page 90.)

Suppose, for example,

$$\frac{d^2 y}{dx^2} = ax^n \ \therefore \ d^2 y = ax^n\,dx^2.$$

Now

$$a \int x^n\,dx = \frac{ax^{n+1}}{n+1} \ \text{and} \int \frac{ax^{n+1}\,dx}{n+1} = \frac{ax^{n+2}}{(n+1)(n+2)},$$

the constants being omitted; hence

$$y = \frac{ax^{n+2}}{(n+1)(n+2)} + C_{,}\,x + C.$$

II.

To integrate the form

$$F\left(y, \frac{d^2 y}{dx^2}\right) = 0.$$

Solving the equation for $\frac{d^2 y}{dx^2}$ we have

$$\frac{d^2 y}{dx^2} = Y \ \therefore \ \frac{dy}{dx} \cdot \frac{d^2 y}{dx^2} = p' \frac{dp'}{dx} = Y \frac{dy}{dx},$$

and multiplying by dx and integrating we have

$$\frac{1}{2} p'^2 = \int Y dy = Y' + C$$

$$\therefore \frac{dy}{dx} = \sqrt{2Y' + 2C} \therefore dx = \frac{dy}{\sqrt{2Y' + 2C}} \therefore x = \int \frac{dy}{\sqrt{2Y' + 2C}} .$$

As an example let

$$a^2 \frac{d^2 y}{dx^2} + y = 0$$

be given to determine the primitive

$$\therefore a^2 \frac{d^2 y}{dx^2} = -y \therefore a^2 p' \frac{dp'}{dx} = -y \frac{dy}{dx},$$

and multiplying by dx and integrating,

$$a^2 p'^2 = -y^2 + C \therefore \frac{dy}{dx} = \frac{\sqrt{C - y^2}}{a}$$

$$\therefore dx = \frac{a dy}{\sqrt{C - y^2}} \therefore x = a \sin.^{-1} \frac{y}{C^{\frac{1}{2}}} + C'.$$

III.

To integrate the form

$$F\left(\frac{dy}{dx}, \frac{d^2 y}{dx^2}\right) = 0.$$

Solving the equation for $\frac{d^2 y}{dx^2}$ as in the preceding cases we have

$$\frac{dp'}{dx} = f p' \therefore dx = \frac{dp'}{f p'}$$

$$\therefore x = \int \frac{dp'}{f p'} \text{ and } y = \int p' dx = \int \frac{p' dp'}{f p'};$$

hence, if p' be eliminated by means of these two equations, the result will be the required relation between x and y.

Let the equation be

$$a \frac{d^2 y}{dx^2} + \left(1 + \frac{dy^2}{dx^2}\right)^{\frac{3}{2}} = 0$$

Y 2

$$\therefore a\,\frac{dp'}{dx} = -\,(1 + \frac{dy^2}{dx^2})^{\frac{3}{2}}$$

$$\therefore dx = \frac{-\,adp'}{(1 + p'^2)^{\frac{3}{2}}} \quad \therefore x = \frac{ap'}{(1 + p'^2)^{\frac{1}{2}}} + C$$

$$dy = \frac{ap'\,dp'}{(1 + p'^2)^{\frac{3}{2}}} \quad \therefore y = \frac{a}{(1 + p'^2)^{\frac{1}{2}}} + C,.$$

The elimination of p' by means of these equations leads to

$$(C - x)^2 + (C, - y)^2 = a^2.$$

This process is obviously the solution of the following problem, viz. To determine the curve whose radius of curvature is constant, for the proposed equation expresses the condition $r = a$.

IV.

To integrate the form

$$F\,(x, \frac{dy}{dx}, \frac{d^2 y}{dx^2}) = 0.$$

Putting $\dfrac{dp'}{dx}$ for its equal $\dfrac{d^2 y}{dx^2}$ the form becomes

$$F\,(x, p', \frac{dp'}{dx}) = 0,$$

which is an equation of the first order between x and p', and of which the integral must be sought for among the methods explained in chapter IV. Supposing this integral to be found and to be

$$f\,(x, p', C) = 0 \ . \ . \ . \ . \ (1),$$

C being the arbitrary constant, then the remainder of the process will depend upon the nature of this equation.

1st. Suppose we can solve it with respect to p', then we may put it under the form

$$p' = X \ \therefore \ dy = X dx \ \therefore \ y = \int X dx,$$

which is the relation between x and y.

2d. Suppose we can solve the equation (1) with respect to x, then, putting P for the resulting function of p', we shall have   .

$$x = P \; \therefore \; x dp' = P dp'.$$

But by integrating by parts the second member of the equation

$$dy = p' \, dx$$

we have

$$y = p' x - \int x dp' \; \therefore \; y = p' x - \int P dp' \; \ldots \; (2);$$

hence if we eliminate p' by means of the equations (1), (2), we shall obtain the sought relation between x and y.

3d. Lastly, if the equation (1) cannot be readily solved for either x or p', then, in order to effect the solution, we must endeavour to integrate (1) by some of the methods taught in chapters III. or IV.

Let the equation be

$$\frac{a^2}{2x} \frac{d^2 y}{dx^2} = (1 + \frac{dy^2}{dx^2})^{\frac{3}{2}} ,$$

this reduces to

$$2x (1 + p'^2)^{\frac{3}{2}} \, dx = a^2 \, dp'$$

$$\therefore \; 2x dx = \frac{a^2 \, dp'}{(1 + p'^2)^{\frac{3}{2}}} ,$$

and integrating this we have

$$x^2 + C = \frac{a^2 p'}{(1 + p'^2)^{\frac{1}{2}}} .$$

This, solved for p' gives

$$p' = \frac{x^2 + C}{\sqrt{a^4 + (x^2 + C)^2}} ,$$

therefore, multiplying by dx, and integrating

$$y = \int \frac{(x^2 + C) \, dx}{\sqrt{a^4 + (x^2 + C)^2}} .$$

Again, let the proposed equation be

$$1 + \frac{dy^2}{dx^2} + x \frac{dy}{dx} \cdot \frac{d^2 y}{dx^2} = a \frac{d^2 y}{dx^2} \sqrt{1 + \frac{dy^2}{dx^2}} .$$

Putting $\dfrac{dp'}{dx}$ for $\dfrac{d^2y}{dx}$, the equation reduces to

$$1 + p'^2 + xp'\,\frac{dy'}{dx} = a\,\frac{dp'}{dx}\,\sqrt{1 + p'^2},$$

or

$$(1 + p'^2)\,dx + xp'\,dp' = a\,(1 + p'^2)^{\frac{1}{2}}\,dp',$$

that is, dividing by $(1 + p'^2)$, we have the form

$$dx + \mathrm{P}x\,dp' = \mathrm{P}'dp',$$

which is a linear equation, the values of P, P′ being

$$\mathrm{P} = \frac{p'}{1 + p'^2}, \quad \mathrm{P}' = \frac{a}{(1 + p'^2)^{\frac{1}{2}}}.$$

Integrating this by the formula at (87), we have

$$x = \frac{ap' + \mathrm{C}}{\sqrt{1 + p'^2}}.$$

Having thus got x, we have from the expression (2), above, viz.

$$y = p'x - \int x\,dp',$$

the value

$$y = p'x - a\,\sqrt{1 + p'^2} - \mathrm{C}\,\log.\,\{p' + \sqrt{1 + p'^2}\} + \mathrm{C}\,\log.\,\mathrm{C}',$$

$$= \frac{\mathrm{C}p' - a}{\sqrt{1 + p'^2}} - \mathrm{C}\,\log.\,\frac{p' + \sqrt{1 + p'^2}}{\mathrm{C},}.$$

It remains, therefore, to eliminate p' by means of these expressions for x and y; the result of this elimination will be found to be

$$y = \sqrt{a^2 + \mathrm{C}^2 - x^2} - \mathrm{C}\,\log.\,\frac{x + a}{\mathrm{C},\,(\mathrm{C} - \sqrt{a^2 + \mathrm{C}^2 - x^2}}$$

which is the required relation between x and y.

Lastly, let the proposed equation be

$$2.\left(a^2\,\frac{dy^2}{dx^2} + x^2\right)\frac{d^2y}{dx^2} = x\,\frac{dy}{dx},$$

which, by making the usual substitution of $\dfrac{dp'}{dx}$ for $\dfrac{d^2y}{dx^2}$ and then multiplying by dx, to prepare it for integration, becomes

$$2\left(a^2 p'^2 + x^2\right) dp' = xp'\, dx.$$

This is a homogeneous equation, and the separation of the variables is effected by substituting $p'z$ for x; whence

$$\frac{dp}{p'} = \frac{zdz}{2a^2 + z^2}$$

$$\therefore \log. p' = \log. \mathrm{C} \sqrt{2a^2 + z^2} \therefore p' = \mathrm{C} \sqrt{2a^2 + z^2}$$

$$\therefore x = p'z = \mathrm{C}z \sqrt{2a^2 + z^2} \ \ldots \ldots (1).$$

To obtain the expression for y in terms of z we need only put in the equation $y = \int p'dx$, the above value for p', and the differential of the last for dx; the result will be

$$y = \frac{2}{3}\, \mathrm{C}^2 z \left(3a^2 + z^2\right) + \mathrm{C}, \ \ldots \ldots (2).$$

The elimination of z by means of the equations (1), (2) will furnish the sought relation between x and y. Or, instead of proceeding in this manner, we may, after substituting $\dfrac{x}{p'}$ for z, in the equation

$$p' = \mathrm{C} \sqrt{2a^2 + z^2},$$

solve the result, viz. the equation

$$p'^2 - \mathrm{C} \sqrt{2a^2 p'^2 + x^2} \text{ or } p'^4 - 2\mathrm{C}a^2 p'^2 = \mathrm{C}x^2$$

for p', we shall thus have $\dfrac{dy}{dx}$ in terms of x, and thence the equation between x and y.

V.

To integrate the form

$$\mathrm{F}\left(y, \frac{dy}{dx}, \frac{d^2y}{dx^2}\right) = 0.$$

Putting as before, $\dfrac{dp'}{dx}$ for its equal $\dfrac{d^2y}{dx^2}$, the form is

$$F\left(y, p'\frac{dp'}{dx}\right)=0,$$

but, because

$$dy = p'dx \therefore dx = \frac{dy}{p'} \therefore \frac{dp'}{dx} = \frac{p'dp'}{dy};$$

hence, by substituting this expression for $\dfrac{dp'}{dx}$, in the above form, we have an equation of the first order among the variables p', y, and their differentials, with which we may proceed as in the former case.

Thus, suppose we had the equation

$$\left(a + y\frac{dy}{dx}\right)\frac{d^2y}{dx^2} = \left(1 + \frac{dy^2}{dx^2}\right)\frac{dy}{dx},$$

then, putting $\dfrac{dp'}{dx}$ for $\dfrac{d^2y}{dx^2}$, we have

$$(a + yp')\frac{dp'}{dx} = (1 + p'^2)\,p',$$

and, putting $p'\dfrac{dp'}{dy}$ for $\dfrac{dp'}{dx}$, and multiplying by $\dfrac{dy}{p'}$, there results

$$(a + yp')\,dp' = (1 + p'^2)\,dy$$

$$\therefore dy - Py\,dp' = P'dp',$$

a linear equation, in which

$$P = \frac{p'}{1 + p'^2}, \; P' = \frac{a}{1 + p'^2};$$

hence (87)

$$y = ap' + C\sqrt{1 + p'^2} \; \cdot \; \cdot \; \cdot \; \cdot \; (1),$$

and therefore

$$dx = \frac{dy}{p'} = a\frac{dp'}{p'} + C\frac{dp'}{\sqrt{1 + p'^2}},$$

consequently

$$x = a\log. p' + C\log. C,(p' + \sqrt{1 + p'^2}). \quad (Ex.\ 7,\ p.\ 34).$$

$$= \log. \{p'^a\,(C,p' + C,\sqrt{1 + p'^2})^C\} \; \cdot \; \cdot \; \cdot \; \cdot \; (2);$$

hence, eliminating p' by means of (1) and (2), the result will be the relation between x and y.

(105.) Besides the foregoing, there are a few other particular cases of the general form (A) that admit of integration, or rather of reduction to forms of the first order; the processes, however, are not only very indirect and embarrassing, but so exceedingly limited in their application, that we shall not hesitate to omit them in this elementary treatise, merely noticing one more case.

And, first, we shall observe, that if we agree to call the coefficient $\dfrac{dy}{dx}$ of 0 dimensions, and the coefficient $\dfrac{d^2y}{dx^2}$ of -1 dimensions, then when, according to this hypothesis, the differential equation of the second order is homogeneous, it may always be reduced to one of the first order by assuming

$$y = vx \text{ and } \frac{d^2y}{dx^2} = \frac{z}{x} \ \ldots \ldots (1).$$

For, let, according to this hypothesis, n be the degree of homogeneity, then it is plain that $\dfrac{d^2y}{dx^2}$ must be multiplied by a factor of $n+1$ dimensions, and as wherever y is, it is to be replaced by vx, it follows that x must enter this factor in $n+1$ dimensions. It is equally plain that x^n will be a factor of $\dfrac{dy}{dx}$; hence, as the other terms rise to the same dimensions, the proposed equation after the substitutions (1) must be divisible by x^n, and the equation will thus be reduced to a function of v, z, p', without x.

Let it be

$$f(v, z, p') = 0 \ \ldots \ldots (2),$$

then since, by hypothesis,

$$dy \text{ or } p'dx = vdx + xdv \ \therefore \ \frac{dx}{x} = \frac{dv}{p'-v} \ \ldots \ldots (3),$$

but

$$dp' \text{ or } \frac{d^2y}{dx^2}dx = \frac{zdx}{x} \ \therefore \ \frac{dx}{x} = \frac{dp'}{z} \ \ldots \ldots (4)$$

$$\therefore \frac{dp'}{z} = \frac{dv}{p' - v} \ \ \dots \ (5).$$

Putting in this last equation for z its value in terms of v, p', as deduced from (2), and the result will obviously be an equation of the first order between v and p', from which p' being determined, and its value in terms of v substituted in (4), we shall have, by integrating,

$$\log. x = \psi v;$$

hence, finally, eliminating v by means of this equation, and the first of (1), above, the result will be the required relation between x and y.

As an example of this process, let us take the equation

$$x \frac{d^2y}{dx^2} - \frac{dy}{dx} = 0,$$

then the substitutions (1) reduce it to

$$z - p' = 0 \ \therefore \ z = p,$$

and this value of z is to be substituted in the equation (5)

$$\therefore \frac{dp'}{p'} = \frac{dv}{p' - v} \ \therefore \ (p' - v) \, dp' = p' \, dx,$$

that is,

$$p'dp' = vdp' + p'dv,$$

each side being an exact differential, we have

$$\frac{1}{2} p'^2 = p'v + C,$$

we are now to determine p' from this, and substitute its value in the equation (4), but because, in the present example, $z = p'$, it will be easier, and amount to the same thing, to determine p' from the equation

$$\frac{dx}{x} = \frac{dp'}{dz} = \frac{dp'}{p'},$$

for we at once get

$$x = Cp' \ \therefore \ p' = \frac{x}{C};$$

this value, substituted in the above integral, gives

$$x^2 = 2Cxv + C_{,,}$$

but

$$y = vx \therefore v = \frac{y}{x}$$

$$\therefore x^2 = 2Cy + C_{,}.$$

(106.) As to equations of a higher order than the second, the general methods of integration are still more limited than those which apply to equations of the second order. There are, however, two classes of equations of the nth order, which may be reduced to the forms of the first and second orders, already integrated.

These forms are

$$F\left(\frac{d^n y}{dx^n} , \frac{d^{n-1} y}{dx^{n-1}}\right) = 0 \cdot \cdot \cdot \cdot (1),$$

and

$$F\left(\frac{d^n y}{dx^n} , \frac{d^{n-2} y}{dx^{n-2}}\right) = 0 \cdot \cdot \cdot \cdot (2).$$

For the first of these let

$$\frac{d^{n-1} y}{dx^{n-1}} = u \therefore \frac{d^n y}{dx^n} = \frac{du}{dx};$$

hence, by substitution, equation (1) becomes

$$F\left(\frac{du}{dx}, u\right) = 0,$$

which is an equation of the first order between u and x; hence, by integrating this, we obtain u in terms of x, that is,

$$u = X \therefore \frac{d^{n-1} y}{dx^{n-1}} = X$$

$$\therefore y = \int^{n-1} X \, dx^{n-1},$$

and the integration may be effected by (53).

In order to integrate the form (2), put

$$\frac{d^{n-2} y}{dx^{n-2}} = u \therefore \frac{d^n y}{dx^n} = \frac{d^2 u}{dx^2};$$

hence, by substitution, the equation (2) becomes

$$F\left(\frac{d^2 u}{dx^2}, u\right) = 0,$$

a form of the second order which we have shewn how to integrate at page 244. Deducing, therefore, the value of u, we have

$$u = X \therefore \frac{d^{n-2}y}{dx^{n-2}} = X \therefore y = \int^{n-2} X \, dx^{n-2}.$$

It will, however, be sometimes convenient to obtain x and dx in terms of u', as well as u in terms of x, as above, because we can readily descend from a coefficient of any order u' to that of the preceding order t', and so on, till we obtain an expression for p' or $\dfrac{dy}{dx}$ in terms of u', in which we may then substitute for u' its value in terms of x, multiply by dx, and integrate.

The following are examples of the foregoing forms:

(107.)   1. Given the differential equation

$$\frac{d^4y}{dx^4} \cdot \frac{d^3y}{dx^3} = 1$$

to determine the complete primitive.

Assume

$$r' = \frac{d^3y}{dx^3} \therefore \frac{dr'}{dx} = \frac{d^4y}{dx^4},$$

so that the proposed is the same as

$$r' \frac{dr'}{dx} = 1 \therefore dx = r'dr' \therefore x = \frac{1}{2} r'^2 + C \therefore r = \sqrt{2(x-C)},$$

as we have got dx in terms of r', we shall be able to deduce the preceding coefficient q' in terms of r' for

$$dq' = r'dx = r'^2 dr' \therefore q' = \frac{r'^3}{3} + C_1$$

$$\therefore dp' = q' \, dx = \{\frac{r'^3}{3} + C_1\} \, rdr \therefore p' = \frac{r'^5}{3 \cdot 5} + \frac{C_1 r'^2}{1 \cdot 2} + C_2$$

$$\therefore dy = p'dx = \{\frac{r'^3}{3 \cdot 5} + C_1 \frac{r'^2}{1 \cdot 2} + C_2\} \, rdr$$

$$\therefore y = \frac{r'^7}{3 \cdot 5 \cdot 7} + C_1 \frac{r'^4}{1 \cdot 2 \cdot 4} + C_2 \frac{r'^2}{1 \cdot 2} + C_3.$$

which is the complete primitive, r being equal to $\sqrt{2(x-C)}$.

2. Given

$$\frac{d^4y}{dx^4} = \frac{d^2y}{dx^2}$$

to determine the complete primitive.

Assume

$$q' = \frac{d^2y}{dx^2} \quad \therefore \quad \frac{d^2q'}{dx^2} = \frac{d^4y}{dx^4},$$

so that the proposed is the same as

$$\frac{d^2q'}{dx^2} = q,$$

from which we get, by integration, (p. 244,)

$$dx = \frac{dq'}{\sqrt{q'^2 + C^2}},$$

whence

$$x = \log. \frac{q' + \sqrt{q'^2 + C^2}}{C_1} \ \ldots \ (1),$$

having got x and dx in terms of q', we may obtain p' in terms of q', thus:

$$dp' = q'dx = \frac{q' \, dq'}{\sqrt{q'^2 + C^2}} \quad \therefore \quad p' = \sqrt{q'^2 + C^2} + C_2$$

$$\therefore \ dy = p'dx = \sqrt{q'^2 + C^2} \, dx + C_2 \, dx = dq' + C_2 \, dx$$

$$\therefore \ y = q' + C_2 \, x + C_3 \ \ldots \ (2);$$

hence we have to eliminate q' by means of (1) and (2); for this purpose put (1) under the form

$$C_1 e^x = q' + \sqrt{q'^2 + C^2}$$

$$\therefore \ C_1^2 e^{2x} - 2q' C_1 e^x = C^2 \therefore \ q' = \frac{C_1^2 e^{2x} - C^2}{2C_1 e^x} = \frac{C_1 e^x}{2} - \frac{Ce^{-x}}{2C_1},$$

therefore, substituting this in (2), we have, for the complete primitive, the form

$$y = ce^x + c_1 e^{-x} + c_2 x + c_3.$$

We shall now pass to the consideration of linear equations.

Linear Equations of the Higher Orders.

(108.) The general form of a linear differential equation of the nth order is

$$\frac{d^n y}{dx^n} + A \, \frac{d^{n-1} y}{dx^{n-1}} + \ldots\ldots + M \frac{dy}{dx} + Ny + X = 0 \ldots (A),$$

A, B, &c. being either functions of x without y, or else constant.

In order to determine the method of integrating this class of equations let us examine a particular case. We shall choose the equation of the third order of the form

$$\frac{d^3 y}{dx^3} + A \, \frac{d^2 y}{dx^2} + B \frac{dy}{dx} + Cy = 0,$$

where x is absent, and in which A, B, C are constants.

Now if we can find a value of y in terms of x, and involving but one arbitrary constant that will satisfy this equation, we know that such an equation between y and x will be a particular case of the complete primitive. The peculiar form of the proposed equation has enabled analysts to discover à priori such a particular case of the primitive, and thence the complete primitive itself. For, from the principles of differentiation, we know that the several differential coefficients derived from an exponential function ce^{mx} all involve this same function, thus

$$y = ce^{mx}, \quad \frac{dy}{dx} = mce^{mx}, \quad \frac{d^2 y}{dx^2} = m^2 ce^{mx}, \quad \frac{d^3 y}{dx^3} = m^3 ce^{mx}, \quad \&c.$$

hence, if this function be put for y, in the proposed equation, all the terms will become divisible by e^{mx}, and the result will be merely an algebraical equation of the third degree in m. But this equation will fix certain values for m, so that, by putting these successively for m in the equation $y = ce^{mx}$, c being a constant, we must have necessarily so many particular values of y which will satisfy the proposed, that is, so many particular cases of the complete primitive. Let us then substitute in the proposed equation

$$y = ce^{mx},$$

which becomes, in consequence,

$$m^3 ce^{mx} + Am^2 ce^{mx} + Bmce^{mx} + Cce^{mx} = 0$$

$$\therefore m^3 + Am^2 + Bm + C = 0 \ldots \ldots (1).$$

Let the three roots of this equation be m_1, m_2, and m_3, then for y we have the three values

$$y_1 = c_1 e^{m_1 x}, \; y_2 = c_2 e^{m_2 x}, \; y_3 = c_3 e^{m_3 x} \ldots \ldots (2),$$

each of which equations necessarily satisfy the proposed. These, therefore, are particular cases of the complete primitive.

As the complete integral must furnish each of these by giving particular values to each of the three arbitrary constants which enter it, this complete integral must be

$$y = c_1 e^{m_1 x} + c_2 e^{m_2 x} + c_3 e^{m_3 x} \ldots \ldots (3).$$

For put successively for y, in the proposed, the values (2), the sum of the results will be

$$
\begin{array}{cccc}
\begin{array}{l} c_1 m_1^3 e^{m_1 x} \\ c_2 m_2^3 e^{m_2 x} \\ c_3 m_3^3 e^{m_3 x} \end{array}
+ A
\begin{array}{l} c_1 m_1^2 e^{m_1 x} \\ c_2 m_2^2 e^{m_2 x} \\ c_3 m_3^2 e^{m_3 x} \end{array}
+ B
\begin{array}{l} c_1 m_1 e^{m_1 x} \\ c_2 m_2 e^{m_2 x} \\ c_3 m_3 e^{m_3 x} \end{array}
+ C
\begin{array}{l} c_1 e^{m_1 x} \\ c_2 e^{m_2 x} = 0, \\ c_3 e^{m_3 x} \end{array}
\end{array}
$$

that is (3)

$$\frac{d^3 y}{dx^3} \quad + \quad A \frac{d^2 y}{dx^2} \quad + \quad B \frac{dy}{dx} \quad + \quad Cy \quad = 0.$$

It is necessary to remark that if any of the roots m_1, m_2, m_3, be equal, as, for instance, $m_1 = m_2$ then (3) will be

$$y = (c_1 + c_2) e^{m_1 x} + c_3 e^{m_3 x},$$

which will not be the complete primitive of the proposed, but only a particular integral, since only two arbitrary constants enter. But when this happens, then it may be shewn that not only is $y = c_1 e^{m_1 x}$, a particular integral of the proposed, but also $y = c_1 x e^{m_1 x}$, for, differentiating this, we have

$$\frac{dy}{dx} = c_1 m_1 x e^{m_1 x} + c_1 e^{m_1 x}, \quad \frac{d^2 y}{dx^2} = c_1 m_1^2 x e^{m_1 x} + 2c_1 m_1 e^{m_1 x}$$

$$\frac{d^3 y}{dx^3} = c_1 m_1^3 x e^{m_1 x} + 3c_1 m_1^2 e^{m_1 x}.$$

Substituting these values in the proposed, we have

$$c_1 x e^{m_1 x} (m_1^3 + A m_1^2 + B m_1 + C) +$$

$$c_1 e^{m_1 x} (3 m_1^2 + 2 A m_1 + B) = 0 \ldots \ldots (4)$$

Now by hypothesis the equation (1) has two equal roots, and it is shown by all writers on the theory of equations that the limiting equation to this, viz.

$$3 m^2 + 2 A m + B = 0$$

has also a root equal to one of these (*See Bridge's Theory of Equations* p. 67). Hence both the terms of (4) vanish, so that the expression (4) is $= 0$, that is, the value $y = c_1 x e^{m_1 x}$ satisfies the proposed equation, and therefore the complete primitive is

$$y = c_1 e^{m_1 x} + c_2 x e^{m_1 x} + c_3 e^{m_3 x}.$$

If all three roots were equal, then it might be shewn, in the same manner, that the complete primitive would be

$$y = c_1 e^{m_1 x} + c_2 x e^{m_1 x} + c_3 x^2 e^{m_1 x}.$$

Let us now consider the linear equation

$$\frac{d^3 y}{dx^3} + A \frac{d^2 y}{dx^2} + B \frac{dy}{dx} + C y + X = 0 \ldots \ldots (B),$$

A, B, C being constants and X a function of x.

Suppose

$$y = C_1 y_1 + C_2 y_2 + C_3 y_3 \ldots \ldots (1),$$

then we know, from what has preceded, that if X were absent from the proposed, that this would be the complete integral of (B) y_1, y_2, y_3 being put for $e^{m_1 x}$, $e^{m_2 x}$, $e^{m_3 x}$ and C , C_2, C_3 being constants. But C_1, C_2, C_3 may be functions of x, and yet of such a nature as to have

no more effect upon the values of $\frac{dy}{dx}$, $\frac{d^2 y}{dx^2}$, $\frac{d^3 y}{dx^3}$, than if they were

constant; for it is only necessary that they be subject to the following three conditions, viz.

$$1. \quad y_1 \frac{dC_1}{dx} + y_2 \frac{dC_2}{dx} + y_3 \frac{dC_3}{dx} = 0,$$

for then

$$\frac{dy}{dx} = C_1 \frac{dy_1}{dx} + C_2 \frac{dy_2}{dx} + C_3 \frac{dy_3}{dx} ,$$

the same as if the coefficients were constant.

$$2. \quad \frac{dC_1\, dy_1}{dx^2} + \frac{dC_2\, dy_2}{dx^2} + \frac{dC_3\, dy_3}{dx^2} = 0,$$

for then

$$\frac{d^2y}{dx^2} = C_1 \frac{d^2y_1}{dx^2} + C_2 \frac{d^2y_3}{dx^2} + C_3 \frac{d^2y_3}{dx^3},$$

the same as for constant coefficients.

Now if, as a third condition, we suppose

$$3. \quad \frac{dC_1\, d^2y}{dx^3} + \frac{dC_2\, d^2y_2}{dx^3} + \frac{dC_3\, d^2y_3}{dx^3} + X = 0,$$

we shall then have

$$\frac{d^3y}{dx^3} = C_1 \frac{d^3y_1}{dx^3} + C_2 \frac{d^3y_2}{dx^3} + C_3 \frac{d^3y_3}{dx^3} - X.$$

Consequently, if we determine C_1, C_2, C_3 from these three conditions, (1) will be the complete primitive of the equation (B).

Such is the theory of Linear differential equations, but for further particulars, and more ample details on this as well as on various other classes of differential equations, which have at different times exercised the powers of analysts, the student must consult works of higher pretensions than the present volume, as *Jephson's Fluxional Calculus*, vol. 2, or the *Calcul Integral* of *Garnier*; but the *Complete Treatise* of *Lacroix*, in three large quarto volumes, furnishes the most extensive view of the labours of analysts in this department of science that has yet appeared.

Determination of Integrals by Approximation.

(109.) The integration of equations of two variables consists in the determination of the general relation between x and y; this determination, however, is, as we have before remarked, not always practicable in finite terms, and in such cases we must content ourselves with an approximation to this relation. The object in view, in the method of approximation, is to determine an expression for one of the variables in a series of ascending or of descending powers of the other, so that, for a proposed numerical value of the one, we may approach to any degree of nearness to the corresponding numerical value of the other. An example or two will shew how these approximations are to be effected.

EXAMPLES.

1. Given the equation

$$\left(1 + \frac{dy}{dx}\right) y = 1$$

to determine y in terms of x.

Assume

$$y = Ax^a + Bx^b + Cx^c + \&c.,$$

where both the exponents and coefficients are indeterminate. By differentiating,

$$\frac{dy}{dx} = Aax^{a-1} + Bbx^{b-1} + Ccx^{-1} + \&c.;$$

hence, by substituting these values of y and $\frac{dy}{dx}$ in the proposed equation, it becomes

$$(1 + Aax^{a-1} + Bbx^{b-1} + Ccx^{c-1} + \&c.)(Ax^a + Bx^b + Cx^c + \&c.) = 1,$$

that is, by actually performing the multiplication here indicated,

$$\left.\begin{array}{l} A^2 ax^{2a-1} + ABax^{a+b-1} + ACax^{a+c-1} + \&c. \\ -1 + ABbx^{a+b-1} + B^2 bx^{2b-1} + \&c. \\ + Ax^a + ACcx^{a+c-1} + \&c. \\ + B^2 x^b + \&c. \end{array}\right\} = 0.$$

We have now so to fix the values of the indeterminate quantities that this equation may hold independently of x, that is, so that the first member may be identically 0; and it is plain that this will be done, provided we can first assume a, b, c, &c. of such values that the exponents may all be equal and can then assume A, B, C, &c., so that the coefficients may mutually destroy each other.

The first object will be accomplished by the conditions

$$2a - 1 = 0,\ a + b - 1 = a,\ a + c - 1 = b,$$

which give

$$a = \frac{1}{2}, \; b = 1, \; c = \frac{3}{2},$$

and for the second we must obviously have the conditions

$$A^2 a = 1, \; AB \, (a + b) + A = 0, \; \&c.$$

which fix for A, B, C, &c. the values

$$A = \sqrt{2}, \; B = -\frac{2}{3}, \; C = \frac{\sqrt{2}}{18}, \; \&c.$$

consequently the required development is

$$y = \sqrt{2} \; x^{\frac{1}{2}} - \frac{2}{3} \; x^{\frac{2}{2}} + \frac{\sqrt{2}}{18} \; x^{\frac{3}{2}} - \&c.$$

It should be remarked that the integral thus determined is not the complete primitive of the proposed, because no arbitrary constant has been introduced. The determination of the arbitrary constant requires that we know the value of y for some given value of x; suppose then that when $x = a$, $y = b$, then the form of the development must be

$$y = b + A \, (x - a)^{\alpha} + B \, (x - a)^{\beta} + \&c. \; \ldots \; (1),$$

and in the present example it is, therefore,

$$y = b + \sqrt{2} \, (x + a)^{\frac{1}{2}} - \frac{2}{3} \, (x - a)^{\frac{3}{2}} + \frac{\sqrt{2}}{18} \, (x - a)^{\frac{3}{2}} - \&c.$$

in which equation the arbitrary constant is involved in a, b. The complete integral is not, however, always so readily determinable; the usual process is to substitute in the proposed differential equation $a + t$ for x and $b + u$ for y, and then to develop u in a series of powers of t, so that when t is made $= 0$, u may become 0, for then when the values of t and u are restored, by the substitution of $x - a$ for t and $y - b$ for u, we shall have $y = b$ when $x = a$ as we ought; or we may at once assume the development of the form (1), and then determine the exponents and coefficients as above.

2. Given the equation

$$\frac{dy}{dx} + y + mx^n = 0$$

to determine the complete integral in a series.

Putting $a + t$ for x, and assuming the development (1), we have

$$y = b + At^{\alpha} + Bt^{\beta} + \&c.$$

$$\therefore \frac{dy}{dt} = A\alpha t^{\alpha-1} + B\beta t^{\beta-1} + C\gamma t^{\gamma-1} + \&c.$$

substituting these values in the proposed, we have, since $\dfrac{dy}{dt} = \dfrac{dy}{dx}$,

$$A\alpha t^{\alpha-1} + B\beta t^{\beta-1} + C\gamma t^{\gamma-1} \qquad\qquad + \&c. +$$

$$b + At^{\alpha} \qquad + Bt^{\beta} \qquad\qquad + \&c. +$$

$$ma^n + mna^{n-1} t + mn\,\frac{n-1}{2}\,a^{n-2}\,t^2 + \&c. = 0.$$

Now to render the exponents the same in the several vertical rows, we must have the conditions

$$\alpha = 1,\ \beta = 2,\ \gamma = 3,\ \&c.;$$

hence

$$A = -ma^n - b,\quad B = \frac{ma^n - mna^{n-1} + b}{1 \cdot 2}$$

$$C = -\frac{ma^n - mna^{n-1} + mn\,(n-1)\,a^{n-2} + b}{1 \cdot 2 \cdot 3}$$

$$\&c. \qquad\qquad\qquad \&c.$$

therefore the exponents and coefficients of the assumed series are determined.

3. Given the equation

$$\frac{d^2 y}{dx^2} + mx^n y = 0$$

to find y in a series.

Assume

$$y = Ax^{\alpha} + Bx^{\beta} + Cx^{\gamma} + \&c.$$

$$\therefore \frac{dy}{dx} = A\alpha x^{\alpha-1} + B\beta x^{\beta-1} + C\gamma x^{\gamma-1} + \&c.$$

$$\therefore \frac{d^2 y}{dx^2} = A\alpha\,(\alpha-1)\,x^{\alpha-2} + B\beta\,(\beta-1)\,x^{\beta-2} + C\gamma\,(\gamma-1)\,x^{\gamma-2}$$

Hence, by substitution, the proposed becomes

$$\left.\begin{array}{l} A\alpha(\alpha-1)x^{\alpha-2} + B\beta(\beta-1)x^{\beta-2} + C\gamma(\gamma-1)x^{\gamma-2} + \&c. \\ \qquad + mAx^{\alpha+n} + mBx^{\beta+n} + mCx^{\gamma+n} + \&c. \end{array}\right\} = 0.$$

To render the exponents the same in the several columns we may suppose $n = -2$, but this would confine the investigation to a particular case of the proposed example. If, however, we first make the term $A\alpha(\alpha-1)x^{\alpha-2}$ vanish by means of the requisite value of α, that is by making either

$$\alpha = 0, \text{ or } \alpha = 1,$$

the above equation will become on the first hypothesis, or $\alpha = 0$,

$$\left.\begin{array}{l} B\beta(\beta-1)x^{\beta-2} + C\gamma(\gamma-1)x^{\gamma-2} + \&c. \\ \qquad + mAx^{n} + mBx^{\beta+n} \qquad\quad + \&c. \end{array}\right\} = 0,$$

consequently, by equalling the exponents,

$$\beta - 2 = n, \ \gamma - 2 = \beta + n, \ \delta - 2 = \gamma + n, \ \&c.$$

$$\therefore \ \beta = n + 2, \ \gamma = 2n + 4, \ \delta = 3n + 6, \ \&c.$$

and equalling the coefficients of the like terms, we get

$$B = \frac{Am}{\beta(\beta-1)}, \ C = \frac{Am^2}{\beta(\beta-1)\gamma(\gamma-1)}, \ \&c.;$$

hence, putting for β, γ, &c. their values just determined and substituting in the assumed series these expressions for the coefficients, we have

$$y = A\left\{1 - \frac{m}{(n+1)(n+2)}x^{n+2} + \right.$$
$$\left. \frac{m^2}{(n+1)(n+2)(2n+3)(2n+4)}x^{2n+4} - \&c.\right\}$$

A being entirely arbitrary.

If we take the second hypothesis, viz. $\alpha = 1$, we shall obtai another expression for y involving an arbitrary constant A, or, for differential sake, A′; this expression will be

$$y = A' \left\{ 1 - \frac{m}{(n+2)(n+3)} x^{n+2} + \frac{m^2}{(n+2)(n+3)(2n+4)(2n+5)} x^{2n+4} - \&c. \right\}$$

the sum of these two particular integrals will be the complete integral of the proposed, involving the two arbitrary constants A, A′, and, by making first one of these 0 and then the other, we have the two particular integrals above deduced.

For other methods of approximation the student may consult the works referred to at the close of last article.

Integration of Simultaneous Equations.

(110.) We shall conclude the present chapter with a few general examples of the integration of simultaneous equations, as they often present themselves in the higher problems of Dynamics.

1. Let it be proposed to integrate the system of equations

$$\left. \begin{array}{l} My + Nx + P\dfrac{dy}{dt} + Q\dfrac{dx}{dt} = T \\[2ex] M'y + N'x + P'\dfrac{dy}{dt} + Q'\dfrac{dx}{dt} = T' \end{array} \right\}$$

which are the most general forms of the first degree between x and y and the differential coefficients $\dfrac{dy}{dt}, \dfrac{dx}{dt}$; and in which M, N, P, &c. are functions of the independent variable t. We may write these equations thus:

$$(My + Nx)\, dt + P\, dy + Q\, dx = T\, dt$$
$$(M'y + N'x)\, dt + P'\, dy + Q'\, dx = T'dt,$$

and if we multiply the second by an indeterminate function θ of t, and add the product to the first, we shall have

$$\{(M + M'\theta)y + (N + N'\theta)x\}\, dt + (P + P'\theta)\, dy + (Q + Q'\theta)\, dx$$
$$= (T + F'\theta)\, dt;$$

that is, putting for brevity

$$M + M'\theta = M_1, \quad N + N'\theta = N_1, \quad P + P'\theta = P_1,$$
$$Q + Q'\theta = Q_1, \quad T + T'\theta = T_1$$

we have the equation

$$M_1\, y dt + N_1\, x dt + P_1\, dy + Q_1\, dx = T_1\, dt,$$

or, which is the same thing,

$$M_1 \left(y + \frac{N_1}{M_1}\, x\right) dt + P_1 \left(dy + \frac{Q_1}{P_1}\, dx\right) = T_1\, dt.$$

Now it is obvious that this equation would agree with the linear equation of the first order, (art. 87,) provided that we had the condition

$$d\left(y + \frac{N_1}{M_1}\, x\right) = dy + \frac{Q_1}{P_1}\, dx \;\dots\; (1),$$

because then by putting

$$y + \frac{N_1}{M_1}\, x = z \;\dots\; (2),$$

the equation becomes, in virtue of the supposed condition (1),

$$M_1\, z dt + P_1\, dz = T_1\, dt,$$

or

$$dz + \frac{M_1}{P_1}\, z dt = \frac{T_1}{P_1}\, dt, \;\dots\; (3),$$

from which equation we know how to obtain z in terms of t, and thence the relation among the variables x, y, and t.

Now to satisfy the condition (1) it is obviously sufficient that we have

$$d\, \frac{N_1}{M_1}\, x = \frac{Q_1}{P_1}\, dx;$$

that is, t being the independent variable,

$$\frac{N_1}{M_1}\, \frac{dx}{dt} + x\, \frac{d\, \dfrac{N_1}{M_1}}{dt} = \frac{Q_1}{P_1}\, \frac{dx}{dt}$$

and as $\dfrac{dx}{dt}$ is indeterminate, the coefficients of this term must be equal, therefore the above condition implies the two

$$\frac{N_1}{M_1} = \frac{Q_1}{P_1}, \quad \frac{d\,\dfrac{N_1}{M_1}}{dt} = 0 \ \ldots \ldots \ (4).$$

If then in these equations we substitute the foregoing values of M_1, N_1, P_1, Q_1, and after having performed the differentiation we eliminate θ, which enters in these equations, the result will be the relation which must subsist among the coefficients of the proposed equations, in order that the integration may depend upon a linear differential equation of the first order. The solution of this linear equation will give z in terms of t, from which we may get y in terms of x and t, and this value of y, substituted in one of the given equations, will furnish a differential equation between x and t, which being integrated we shall finally obtain the values of x and y in terms of t.

When the coefficients M, N, P, &c. in the first members of the proposed equations are all constant, the second condition (4) is necessarily satisfied, when θ is constant, and we shall then have only to determine the arbitrary factor θ, so that the other condition may have place. This first condition, by restoring the values of M_1, N_1, P_1, Q_1, is

$$\frac{N + N'\theta}{M + M'\theta} = \frac{Q + Q'\theta}{P + P'\theta},$$

which, by reducing to a common denominator, furnishes a quadratic equation in θ. Let its roots be θ' and θ'', and the corresponding values of the coefficients of (3), m and n in the first case, and m' and n' in the second, then the equation (3) gives the two

$$dz + m\,z\,dt = n\,dt$$

$$dz + m'z\,dt = n'dt,$$

and these integrated by the formula at page 192 furnish the two equations

$$z = e^{-\int m\,dt}\left\{\int n\,e^{\int m\,dt}\,dt\right\}$$

$$z = e^{-\int m'dt}\left\{\int n'e^{\int m'dt}\,dt\right\};$$

hence, putting in these the value of z (2), we shall have two equations in x, y, and t, from which both x and y may be obtained in terms of t.

2. Let it now be required to integrate the system

$$dy + (M\,y + N\,x + P\,z)\,dt = T\,dt$$
$$dx + (M\,y + N'x + P'z)\,dt = T'\,dt$$
$$dz + (M''y + N''x + P''z)\,dt = T''dt$$

in which all the coefficients are constant except T, T', T", which are functions of the independent variable t.

Multiplying the second by a constant C, and the third by another constant C', and adding the products to the first, we shall have an equation of the form

$$dy + Cdx + C'dy + Q\,(y + Rx + Sz)\,dt = Udt,$$

which, as in the former case, will agree with a linear differential equation, provided we have the condition

$$d\,(y + Rx + Sz) = dy + Cdx + C'dy,$$

which requires that

$$C = R, \quad C' = S;$$

hence, as C and C' are contained in R and S, these two equations will suffice to determine the different values of C, C', which will cause the required condition to exist; or which will render the proposed equations integrable by means of linear equations of the first order.

The above method applies to differential equations of the superior orders, because these may be reduced to equations of the first order. Thus, for example, if the equations were

$$d^2y + (My + N\,x)\,dt^2 + (P\,dy + Q\,dx)\,dt = T\,dt^2$$
$$d^2x + (M'y + N'x)\,dt^2 + (P'dy + Q'dx)\,dt = T'dt^2$$

we should be able to reduce them to four equations of the first order, viz.

$$dy = p'\,dt, \quad dx = q'\,dt$$
$$dp' + (My + N\,x + Pp' + Q\,q')\,dt = T\,dt$$
$$dq' + (M'y + N'x + P'p' + Q'q')\,dt = T'dt$$

and to these four equations the preceding process may be applied For particular examples of the integration of simultaneous differential equations, we must refer to Peacock's Collection of Examples.

CHAPTER VII.

ON THE INTEGRATION OF TOTAL DIFFERENTIAL EQUATIONS OF THREE VARIABLES.

(110.) Let

$$P dx + Q dy + R dz = 0 \ldots (1)$$

be a differential equation of three variables, of which the two x and y are entirely independent. By putting

$$p = - \frac{P}{R}, \; q = - \frac{Q}{R} \ldots (2)$$

this equation becomes

$$dz = p dx + q dy \ldots (3).$$

If this is the *total* differential of z, immediately derivable from some primitive

$$z = F(x, y) \ldots (4),$$

then we know that we must have

$$\frac{dz}{dx} = p, \; \frac{dz}{dy} = q \ldots (5),$$

and, moreover, that the second member of (3) must fulfil Euler's condition of integrability (78): for although z may enter p and q as well as x and y, yet as z is a function of x and y the second member of (3) is a differential expression of but two independent variables. The condition of integrability is, therefore,

$$\{\frac{dp}{dy}\} = \{\frac{dq}{dx}\},$$

that is,

$$\frac{dp}{dy} + \frac{dp}{dz} \frac{dz}{dy} = \frac{dq}{dx} + \frac{dq}{dz} \frac{dz}{dx}.$$

By transposing we have, in virtue of (5),

$$\frac{dp}{dy} - \frac{dq}{dx} + q \frac{dp}{dz} - p \frac{dq}{dz} = 0 \ldots (6),$$

which expresses the condition of integrability. But to have this condition in terms of P, Q, R, instead of p and q, we have, by differentiating (2),

$$\frac{dp}{dy} = \frac{R\,\dfrac{dP}{dy} - P\,\dfrac{dR}{dy}}{R^2}, \quad \frac{dq}{dx} = -\frac{R\,\dfrac{dQ}{dx} - Q\,\dfrac{dR}{dx}}{R^2}$$

$$q\frac{dp}{dz} = \frac{Q}{R}\;\frac{R\,\dfrac{dP}{dz} - P\,\dfrac{dR}{dz}}{R^2}, \quad p\frac{dq}{dz} = \frac{P}{R}\cdot\frac{R\,\dfrac{dQ}{dz} - Q\,\dfrac{dR}{dz}}{R^2};$$

hence, by substituting these values in (6), the equation of condition reduces to

$$R\frac{dP}{dy} - P\frac{dR}{dy} - R\frac{dQ}{dx} + Q\frac{dR}{dx} - Q\frac{dP}{dz} +$$

$$P\frac{dQ}{dz} = 0 \ldots \ldots (7),$$

and which equation must exist if the equation (4) exists; that is, if there can exist an equation among the three variables x, y, z, in conjunction with (1). Consequently, if we take at hazard a differential equation

$$Mdx + Ndy + Pdz = 0,$$

then, without first ascertaining whether the condition (7) exists, we cannot affirm that one of the three variables is a function of the other two, considered as independent, or that this differential equation necessarily implies the existence of some equation among x, y, z. Formerly, however, those differential equations which did not fulfil the condition (7) were considered to be meaningless, but *Monge* proved this supposition to be erroneous, and shewed that although to such equations there corresponded no single primitive, yet they might be satisfied by a pair of primitive equations involving an arbitrary function of the dependent variable z, their geometrical signification being an infinite variety of curves of double curvature: we shall advert to this presently.

It must be remarked, that the existence of the condition (6) or (7) does not imply that the proposed (1) is an exact differential, although it does imply that (1) is integrable, for, otherwise, the second member of (3) could not be an exact differential, which it is by hypothesis; but this second member, it is easy to see, remains unaltered by whatever factor we multiply (1), so that when we have ascertained that the condition (7) has place for any proposed differential, we must, in order to integrate it, determine the factor, which will render it exact.

Let us then suppose that the differential

$$M dx + N dy + P dz = 0 \ldots \ldots (8),$$

will become the immediate differential of some function of x, y, z, represented by $U = 0$, upon being multiplied by the factor λ, then, for the total differential of U, we have

$$dU = M\lambda \, dx + N\lambda \, dy + P\lambda \, dz = 0.$$

Now as z enters into the two first terms of this complete differential the same as if it were a constant, we shall obtain the integral U by integrating the equation

$$M\lambda \, dx + N\lambda \, dy = 0 \ldots \ldots (9),$$

N and N being functions of the variables x, y and of the constant z, provided we determine the arbitrary constant, which may obviously be a function of the constant z, so that the complete integral may be the same as U. Representing, then, the complete integral of (9) by

$$U = V + \phi z = 0,$$

it will remain to determine ϕz. For this purpose let us differentiate with respect to z, and we ought to have

$$\frac{dU}{dz} = \frac{dV}{dz} + \frac{d\phi z}{dz} = P\lambda$$

$$\therefore \quad \frac{d\phi z}{dz} = P\lambda - \frac{dV}{dz}$$

$$\therefore \quad \phi z = \int \left(P\lambda - \frac{dV}{dz} \right) dz,$$

and thus the function ϕz becomes known.

Since ϕz contains neither of the variables x, y, they must both be absent from

$$P\lambda - \frac{dV}{dx} \, ;$$

if, therefore, either of them were to enter this expression, we must infer that the factor λ has not been properly chosen, for, although it render (9) integrable, it will not in this case render (8) so.

It is obvious that the factor which renders (8) integrable, renders not only (9) but also the two other partial equations

$$\mathrm{M}\lambda\, dx + \mathrm{P}\lambda\, dz = 0$$

$$\mathrm{N}\lambda\, dy + \mathrm{P}\lambda\, dz = 0$$

integrable, the factor, therefore, must be chosen so as to fulfil these three conditions.

(111.) As an example, let the proposed equation be

$$yz\, dx - xz\, dy + yx\, dz = 0.$$

This satisfies the equation (7), it is, therefore, integrable, and to ascertain whether any and what factor is necessary to render it an exact differential, we must first consider one of the variables as z constant, writing the equation thus:

$$z\,(ydx - xdy) = 0 \;\ldots\; (1),$$

this does not satisfy the condition of integrability, but (94) it is rendered exact by the factor $\lambda = \dfrac{1}{y^2}$, and this same factor is found to render also the other two partial equations exact. Multiplying then (1) by $\dfrac{1}{y^2}$ and integrating, we have, omitting the constant,

$$\mathrm{U} = \frac{zx}{y} + \phi z = 0 \;\ldots\; (2)$$

$$\therefore \frac{d\mathrm{U}}{dz} = \frac{x}{y} \;\Big|\; \frac{d\phi z}{dz} = yx \cdot \frac{1}{y^2} = \frac{x}{y}$$

$$\therefore \frac{d\phi z}{dz} = 0 \;\therefore\; \phi z = \mathrm{C}\,;$$

hence, substituting this value in (2), we have for the sought integral

$$\mathrm{U} = \frac{zx}{y} + \mathrm{C}.$$

Again, let the equation

$$zy\,dx + xz\,dy + xy\,dz + az^2\,dz = 0$$

be proposed. This also satisfies the condition (7); we shall, therefore, first integrate

$$z\,(y\,dx + x\,dy) = 0$$

on the hypothesis of z constant, and we find for the integral

$$U = zxy + \phi z = 0 \ \ldots \ (1),$$

so that no factor is here requisite,

$$\therefore \frac{dU}{dz} = xy + \frac{d\phi z}{dz} = xy + az^2$$

$$\therefore \frac{d\phi z}{dz} = az^2$$

$$\therefore \phi z = a \int z^2\,dz = \frac{az^3}{3} + C;$$

hence, by substitution, the integral U becomes

$$U = zxy + \frac{az^3}{3} + C.$$

(112.) Let us now consider the case in which the differential equation

$$Mdx + Ndy + Pdz = 0 \ \ldots \ (1)$$

does not satisfy the condition (7), and let λ be the factor proper to render integrable the part $Mdx + Ndy$ only, z being regarded as constant; by multiplying the proposed by this factor, it becomes

$$M\lambda dx + N\lambda dy + P\lambda dz = 0 \ \ldots \ (2).$$

Integrating the equation

$$M\lambda dx + N\lambda dy = 0 \ \ldots \ (3)$$

we have, as before,

$$V + \phi z = 0 \ \ldots \ (4),$$

but the differential of this equation, taken with respect to the three variables, cannot, as in the former case, be identical with (2), which it would however be, if its differential with respect to z were equal to $P\lambda$. Now the differential of (4) with respect to the three variables is, in virtue of (3),

$$M\lambda dx + N\lambda dy + \frac{dV}{dz}\,dz + \frac{d\phi z}{dz}\,dz = 0 \ldots \ldots (5).$$

and its differential with respect to z only is

$$\left(\frac{dV}{dz} + \frac{d\phi z}{dz}\right)dz,$$

and therefore, although it is impossible that any equation (4) among the variables x, y, z can be found, whose differential (5) shall be identical to (3), without assuming some other relation among the variables, yet, by introducing a new relation, viz. the relation

$$\frac{dV}{dz} + \frac{d\lambda z}{dz} = P\lambda,$$

the identity is brought about, for, in virtue of this condition, (5) becomes

$$M\lambda dx + N\lambda dy + P\lambda dz = 0,$$

and thus the proposed differential equation is satisfied by the equations

$$\left.\begin{aligned}V + \phi z &= 0 \\ \frac{dV}{dz} + \frac{d\phi z}{dz} &= P\lambda\end{aligned}\right\} \ldots \ldots (6)$$

taken conjointly, in which the function ϕz is entirely arbitrary. The system of equations (6) involving an arbitrary function of z represents an infinite variety of curves of double curvature, all of which equally give rise to the differential equation (1) or (2).

Suppose, for example,

$$ydy + zdx - dz = 0,$$

an equation which does not satisfy the condition (7).

Regarding z as constant, the factor necessary to render the part

$$ydy + zdx$$

integrable is 2, consequently the proposed, multiplied by this, is

$$2ydy + 2zdx - 2dz = 0,$$

which equation is satisfied by the system of equations

$$y^2 + 2zx + \phi z = 0 \left.\right\}$$
$$2a + \frac{d\phi z}{dz} + 2 = 0 \left.\right\}$$

If we take $\phi z = z^3$, the system is

$$y^2 + 2zx + z^3 = 0 \left.\right\}$$
$$2x + 3z^2 + 2 = 0 \left.\right\}$$

and so on.

CHAPTER VIII.

ON THE INTEGRATION OF PARTIAL DIFFERENTIAL EQUATIONS.

Partial Differential Equations of the First Order.

(113.) A partial differential equation of the first order, containing three variables x, y, z, is one which, besides the variables themselves and constant quantities, contains only the partial differential coefficients $\frac{dz}{dx}$, $\frac{dz}{dy}$. The integration of this class of differential equations forms a distinct and very extensive branch of the calculus, involving difficulties of a peculiar kind. In the present small volume we must confine ourselves to a very elementary view of the subject, referring the student for further information to the large work of *Lacroix*, before mentioned.

I.

To integrate the partial differential equation

$$\frac{dz}{dx} = \mathrm{X},$$

X being a function of x, and z a function of the independent variables x, y.

Multiplying by dx, and integrating, we have

$$z = f(x, y) = \int X dx + \phi y,$$

the arbitrary function ϕy supplying the place of the arbitrary constant, because the partial differential coefficient $\dfrac{dz}{dx}$ has been deduced from the hypothesis of y constant.

Suppose, for example, the equation were

$$\frac{dz}{dx} = x^2 + a^2,$$

then

$$z = \frac{x^3}{3} + a^2 x + \phi y.$$

II.

To integrate the partial differential equation

$$\frac{dz}{dx} = P,$$

P being a function of x, y and z.

As the coefficient $\dfrac{dz}{dx}$ is deduced on the hypothesis of y constant, we must preserve this hypothesis in returning to the original function $z = f(x, y)$; hence, multiplying by dx, we have

$$dz = P dx,$$

the y in P being considered as a constant; this will be a differential equation between the two variables z, x, the integral of which must be completed by annexing the arbitrary function ϕy.

EXAMPLES.

1. Let the equation be

$$\frac{dz}{dx} = \frac{x}{\sqrt{x^2 + y^2}}$$

$$\therefore z = \sqrt{x^2 + y^2} + \phi y.$$

2. Let the equation be

$$\frac{dz}{dx} = \frac{a}{\sqrt{x^2 - y^2 - x^2}}$$

$$\therefore z = a \sin.^{-1} \frac{x}{\sqrt{a^2 - y^2}} + \phi y.$$

3. Let the equation be

$$z \frac{dz}{dx} = \sqrt{y^2 - z^2}$$

$$\therefore \frac{z dz}{\sqrt{y^2 - z^2}} = dx,$$

therefore, y being considered constant,

$$- \sqrt{y^2 - z^2} = x + \phi y,$$

or

$$x + \sqrt{y^2 - z^2} = \phi y.$$

4. Let the equation be

$$\frac{dz}{dx} = \frac{y^2 + z^2}{y^2 + x^2}$$

$$\therefore \frac{dz}{y^2 + z^2} = \frac{dx}{y^2 + x^2},$$

and, integrating on the hypothesis that y is constant, we have

$$\frac{1}{y} \tan.^{-1} \frac{z}{y} = \frac{1}{y} \tan.^{-1} \frac{x}{y} + \phi y.$$

III.

To integrate the partial differential equation

$$M \frac{dz}{dy} + N \frac{dz}{dx} = 0,$$

M and N being functions of x and y.

From this equation, we get

$$\frac{dz}{dy} = - \frac{M}{N} \frac{dz}{dx},$$

and since, by hypothesis, z is a function of x and y,

$$\therefore \ dz = \frac{dz}{dx} dx + \frac{dz}{dy} dy,$$

and we have, by substitution,

$$dz = \frac{dz}{dx} \left\{ dz - \frac{M}{N} dy \right\},$$

or

$$dz = \frac{dz}{dx} \cdot \frac{Ndx - Mdy}{N}.$$

Suppose λ is the factor which renders $Ndx - Mdy$ an exact differential du, that is, let

$$\lambda \left(Ndx - Mdy \right) = du,$$

then the preceding equation becomes

$$dz = \frac{1}{\lambda N} \cdot \frac{dz}{dx} du,$$

to satisfy which we need only assume

$$\frac{1}{\lambda N} \cdot \frac{dz}{dx} = Fu,$$

for then

$$dz = Fu \cdot du \ \therefore \ z = \phi u,$$

ϕ being entirely arbitrary, and u a known function of x and y.

1. Let the equation be

$$x \frac{dz}{dy} - y \frac{dz}{dx} = 0,$$

which is the general partial differential equation of surfaces of revolution (*Diff. Calc.* p. 175).

In this case

$$N dx - M dy = x dx + y dy,$$

which is rendered integrable by $\lambda = 2$,

$$\therefore u = x^2 + y^2,$$

and, consequently,

$$z = \phi (x^2 + y^2),$$

the general equation of surfaces of revolution.

As a second example, let the partial differential equation

$$x \frac{dz}{dx} + y \frac{dz}{dy} = 0$$

be proposed, which belongs to right conoidal surfaces in general, then

$$N dx - M dy = y dx - x dy,$$

which is rendered integrable by $\lambda = \dfrac{1}{y^2}$: hence

$$u = \int \frac{y dx - x dy}{y^2} = \frac{x}{y}$$

$$\therefore z = \phi \frac{x}{y}$$

an equation which we know is the general representation of all right conoidal surfaces (*Diff. Calc.* p. 199-201).

IV.

Let now the form

$$P \frac{dz}{dx} + Q \frac{dz}{dy} + R = 0$$

be proposed in which P, Q, R are functions of x, y, and z; then, dividing by P, and putting

$$\frac{Q}{P} = M, \frac{R}{P} = N,$$

the form becomes

$$\frac{dz}{dx} + M\frac{dz}{dy} + N = 0,$$

that is, putting p' for $\dfrac{dz}{dx}$, and q' for $\dfrac{dz}{dy}$,

$$p' + Mq' + N = 0 \ldots \ldots (A).$$

As this equation exists in conjunction with

$$dz = p'dx + q'dy,$$

which merely implies that z is a function of x and y, we may eliminate p', and we shall thus have

$$dz + Ndx = q'(dy - Mdx) \ldots \ldots (B),$$

this equation being true, whatever be the value of q', we must have separately

$$dz + Ndx = 0, \quad dy - Mdx = 0^* \ldots \ldots (1).$$

Now, if it should so happen that z is absent from both N and M, then the second equation will imply some relation between x and y, furnished by the integral of that equation. Supposing then λ to be the factor which renders it an exact differential, we shall have

$$\lambda(dy - Mdx) = 0,$$

and, by integrating, we get an equation of the form

---

* It may be proper to remark here that these equations, in their present form, teach us nothing, since, from the first principles of the calculus, we know that dx, dy, and dz are necessarily each 0. They are, however, immediately reducible to a significant form, by dividing by dx, since they then become

$$\frac{dx}{dx} + N = 0, \quad \frac{dy}{dx} - M = 0,$$

in which latter equation it must be observed that although $\dfrac{dy}{dx}$ implies a relation between x and y, yet as we may consider this relation to be arbitrary, we shall in effect consider x and y to be independent.

$$F(x, y) = C \quad . \quad . \quad . \quad . \quad (2)$$

$$\therefore y = f(x, C);$$

consequently, substituting this value of y, in the function N, we shall have

$$z = - \int N dx,$$

the second member of this equation being a function of x, and of the constant C, in which, after integrating, if we restore the value of C (2) the result will be the sought relation among the three variables, taking care, however, to consider the arbitrary constant which completes the integral to be an arbitrary function of the constant C, in order that when the value of C (2) is restored, the integral may not be deficient in generality.

As an example, let the equation be

$$x \frac{dz}{dx} + y \frac{dz}{dy} = a \sqrt{x^2 + y^2}.$$

Comparing it with (A), we have

$$M = \frac{y}{x}, \; N = - a \frac{\sqrt{x^2 + y^2}}{x};$$

hence the two equations (1) are

$$dz - a \frac{\sqrt{x^2 + y^2}}{x} dx = 0, \; dy - \frac{y}{x} dx = 0 \quad . \quad . \quad . \quad . \quad (3),$$

z being absent from each.

Now the factor λ, which renders the second of these equations, or rather the equation

$$\frac{x dy - y dx}{x} = 0,$$

integrable, is $\lambda = \frac{1}{x}$; multiplying then by this, and integrating, we have

$$\frac{y}{x} = C \; \therefore y = Cx,$$

consequently, the first of these becomes

$$dz = adx \sqrt{1 + C^2}$$

$$\therefore z = ax \sqrt{1 + C^2} + \phi C,$$

where ϕC may, of course, contain another arbitrary constant besides C. Restoring, now, the value of C, we obtain, finally,

$$z = ax \sqrt{1 + \frac{y^2}{x^2}} + \phi \frac{y}{x}, \quad \cdots$$

or rather

$$z = a \sqrt{x^2 + y^2} + \phi \frac{y}{x} \cdot \cdot \cdot \cdot (4),$$

ϕ being a function quite arbitrary.

If, by differentiation, we eliminate the arbitrary function ϕ, (see *Diff. Calc.* p. 82), we shall return to the original partial differential equation. If $a = 0$, in the proposed, it will be the general representation of right conoidal surfaces, before noticed, and the equation (4) will be the integral equation of the same class of surfaces.

It may happen that the two members of the equation (B) may contain each only the variables whose differentials are involved in them, so that y may be absent from N, and z from M. In this case let, as before, λ be the factor which renders the expression $dy - Mdx$ integrable, and let λ' be the factor which renders $dz + Ndx$ integrable, the members of the equation (B) may then be represented by

$$dz + Ndx = \frac{1}{\lambda'} dU, \quad dy - Mdx = \frac{1}{\lambda} dV,$$

so that we shall have

$$dU = q' \frac{\lambda'}{\lambda} dV \cdot \cdot \cdot \cdot (5),$$

the first member of this equation is an exact differential, and that the second member may be also exact, we must have

$$q' \frac{\lambda'}{\lambda} = \phi V,$$

which is the only condition which need restrict the arbitrary function q'; hence, by substitution, in (5)

$$U = \Phi V,$$

that is to say, U is an arbitrary function of V, U and V being functions of the variables already determined.

Let, for example, the equation

$$xy\frac{dz}{dx} + x^2\frac{dz}{dy} = yz$$

be proposed, which will accord with the general equation (A), if writtten thus:

$$\frac{dz}{dx} + \frac{x}{y}\frac{dz}{dy} - \frac{z}{x} = 0,$$

M and N being

$$M = \frac{x}{y}, N = -\frac{z}{x}.$$

hence the equation (B) arising from the elimination of p', is

$$dz - \frac{z}{x}dx = q'\left(dy - \frac{x}{y}dx\right),$$

so that we have now to find factors which shall render integrable the expressions

$$dz - \frac{z}{x}dx, \; dy - \frac{x}{y}dx,$$

these factors are $\dfrac{1}{x}$ and $2y$; multiplying, therefore, by these, we have the exact differentials

$$\frac{xdz - zdx}{x^2}, \; 2y\,dy - 2x\,dx,$$

of which the integrals are

$$U = \frac{z}{x}, \; V = y^2 - x^2,$$

consequently the required integral is

$$\frac{z}{x} = \Phi(y^2 - x^2).$$

It should be here remarked, that instead of eliminating p' from the equation (A), as we have done in the preceding examples, we may eliminate q', and deal with the resulting equation

$$Mdz + Ndy = q'\,(dy - Mdx) = 0,$$

as we have already dealt with (B).

On the Determination of Arbitrary Functions.

(115.) In all the preceding examples of the integration of partial differential equations, the integral involves an arbitrary function of some of the variables, which ought to be the case, since, as shewn in the Differential Calculus, p. 82, any arbitrary function involved in an integral equation may be eliminated by differentiation, and the resulting equation will always be a partial differential equation of the first order. This elimination was very frequently performed in our section on the *Theory of Curve Surfaces*. In returning, therefore, from the partial differential equation to the original primitive, this last, to be perfectly general, ought to involve an arbitrary function, in the same manner as the integrals of ordinary differential equations involve an arbitrary constant. We know that in this latter class of equations the deter mination of the arbitrary constant, in any particular case, depends upon the nature and conditions of the problem to which it applies; and so also with respect to the arbitrary functions which supply the place of these constants in the former class of equations, their determination depends on the nature of the particular problems to which they belong. For example, the primitive of the equation

$$a\,\frac{dz}{dx} + b\,\frac{dz}{dy} = 1 \;\ldots\;(2),$$

is

$$y - bz = \phi\,(x - az) \;\ldots\;(2),$$

which represents cylindrical surfaces in general, without regard to the nature of the directrix. But if we knew, from the conditions of the problem, the equation, $y = fx$, of the directrix of the particular cylinder which is the subject of inquiry, then, although the differential equation (2) would remain unrestricted, since nothing arbitrary is involved in it, yet its integral (2) would be restricted by this condition, viz. that when $z = 0$, the equations (2) and $y = fz$ are identical, because (2) will then represent the trace of the cylinder on the plane of xy, that is, the directrix; hence the condition is that

$$\phi x = fx,$$

so that ϕ remains no longer arbitrary, but becomes the known form f, therefore the particular integral corresponding to the particular cylinder in question is

$$y - bz = f(x - az),$$

$y - bz$ being substituted for y, and $x - az$ for x, in the given equation of the directrix (*See Anal. Geom.* p. 244).

If the given directrix were a curve of double curvature, then, putting

$$x - az = u,$$

we may, by means of this equation and the two given equations of the directrix, determine the values of x, y, and z, in terms of u, and consequently $y - bz$, or its equal ϕu, will be determined in terms of u, or in other words the form of ϕ will become known, and in which we shall then have merely to put $x - az$ for u, to have the equation of the particular cylindrical surface which is the object of inquiry.

Again the primitive of the partial differential equation

$$\frac{dz}{dx} x + \frac{dz}{dy} = 0$$

is

$$z = \phi\left(\frac{y}{x}\right) \text{ or } \phi^{-1} z = \frac{y}{x},$$

which belongs to every conoidal surface whose straight directrix coincides with the axis of z, without any regard to the nature of the curvilinear directrix; but if this is fixed by the conditions of the problem, then, putting $u = z$, we may, by means of this equation, and those of the given directrix, determine the values of x, y, and z, in terms of u, consequently $\phi^{-1} z$ will be determined in terms of u or of z, so that the form of this function will become known, and thus the particular equation sought will be determined.

Should, however, the problem in question furnish no conditions for the determination of the arbitrary function, then the geometrical representation of the integral comprises an infinity of surfaces, not, however altogether arbitrary, but entirely arbitrary as far as depends upon the arbitrary function. For example, the primitive of the partial differential equation

$$\frac{dz}{dx} = a \ \ldots \ (1)$$

is

$$z = ax + \phi y \ \ldots \ (2),$$

an equation which represents an infinite variety of surfaces according to the infinite variety of arbitrary forms we give to ϕy; but yet all these surfaces must possess this common property, indicated by (1), viz. that if each be cut by a plane parallel to that of xz, the inclination of the section at any point (x, z) must be constant, and equal to a, (*Diff. Calc.* p. 162), consequently every such section must be a straight line; thus far, therefore, the surfaces comprised in (2) are restricted. If, in (2), we suppose $x = 0$, then we shall have, for the trace of any of the surfaces on the plane of xy, the equation

$$z = \phi y,$$

which is entirely unrestricted, so that no curve can be even conceived which this equation shall not comprise; for even if the curve be described at random, since each point in it will be comprised in this equation, their locus will be comprised in it.

What has been said in the present chapter on the subject of partial differential equations, and on the arbitrary constants which their integrals involve, is intended to convey only a few elementary notions of a very extensive and very difficult department of analysis: the full development of the theory of partial differential equations is what cannot be expected in an elementary volume like the present: we hope, however, to return to this subject at some future opportunity.

THE END OF THE INTEGRAL CALCULUS.

NOTES.

NOTES.

Note (A), *page 35.*

(Supplement to. Chapter II.*)*

We have fully explained in the text the method of finding, by indeterminate coefficients, the numerators A, B, C, &c. of the several partial fractions into which any rational fraction may be decomposed: we propose here to shew how the same numerators may be determined by the application of the differential calculus.

1. Let us first consider those partial fractions which arise from the real roots of the denominator of the proposed. If m of these roots are equal, we know (12) that the partial fractions to which the factor $(x-a)^m$ involving these roots gives rise, are

$$\frac{A}{(x-a)^m} + \frac{B}{(x-a)^{m-1}} + \cdots \cdot \frac{K}{x-a} \cdots (1),$$

which if $m=1$ becomes simply $\dfrac{A}{x-a}$.

Hence if $\dfrac{U_1}{V_1}$ be put to represent the sum of the remaining partial fractions which make up the proposed $\dfrac{U}{V}$ then we have

$$\frac{U}{V} = \frac{A}{(x-a)^m} + \frac{B}{(x-a)^{m-1}} + \cdots \cdot \frac{K}{x-a} +$$

$$\frac{U_1}{V_1} = \frac{U}{V_1(x-a)^m} \cdots (2).$$

Multiplying the second and third members by V_1 we get for U_1 the expression

$$U_1 = \frac{V_1\left\{\dfrac{U}{V_1} - A - B(x-a) - C(x-a)^2 \ldots - K(x-a)^{m-1}\right\}}{(x-a)^m} \ldots(3).$$

Now, since V_1 is not divisible by $(x-a)^m$, this expression informs us that the quantity within the brackets must be divisible by $(x-a)^m$, so that this quantity must be of the form $X(x-a)^m$, X being a rational function of x. It follows, therefore, that if we differentiate successively this same quantity, each of the coefficients, from the first to the $(m-1)$th, must be equal to 0 when a is substituted for x: in virtue of this property we shall be readily enabled to determine the numeral coefficients in the numerator of (3). For, in the first place, it is plain from (2) that by multiplying the second and third members by $(x-a)^m$ and then putting $x = a$ we shall obtain the value of A, viz.

$$A = \left[\frac{U}{V_1}\right],$$

the brackets being intended to intimate that a particular value is given to x, viz. $x = a$; differentiating now the expression within the brackets (3), we have, in virtue of the property just established,

$$\left[\frac{d\dfrac{U}{V_1}}{dx}\right] = B, \quad \left[\frac{d^2\dfrac{U}{V_1}}{dx^2}\right] = 2C, \quad \left[\frac{d^3\dfrac{U}{V_1}}{dx^3}\right] = 2 \cdot 3D, \ \&c.$$

hence

$$A = \left[\frac{U}{V_1}\right], \ B = \left[\frac{d\dfrac{U}{V_1}}{dx}\right], \ C = \frac{1}{2}\left[\frac{d^2\dfrac{U}{V_1}}{dx^2}\right], \ D = \frac{1}{2\cdot 3}\left[\frac{d^3\dfrac{U}{V_1}}{dx^3}\right], \ \&c.$$

in this way therefore the partial fractions (1) may be determined.

2. Let us now proceed to determine the partial fractions corresponding to imaginary roots. In this case (13),

$$\frac{U}{V} = \frac{Ax+B}{\{(x-a)^2+\beta^2\}^m} + \frac{Cx+D}{\{(x-a)^2+\beta^2\}^{m-1}} \cdots +$$

$$\frac{Ix+K}{\{(x-a)^2+\beta^2\}^{m-2}} + \frac{U_1}{V_1} = \frac{U}{V_1\{(x-a)^2+\beta^2\}^m}.$$

Multiplying the second and third members by V_1 we get for U_1 the expression.

$$U_1 = \frac{V_1 \left\{ \dfrac{U}{V_1} - (Ax + B) - (Cx + D) \left\{ (x - a)^2 + \beta^2 \right\} - \dots \right\}}{\left\{ (x - a)^2 + \beta^2 \right\}^m}.$$

As before, the expression within the brackets must be divisible by the denominator, and must, therefore, be of the form

$$X \left\{ (x - a)^2 + \beta^2 \right\}^m;$$

hence the successive differential coefficients, from the first to the $(m - 1)$th, become each equal to 0 when for x is substituted one of the roots of

$$(x - a)^2 + \beta^2 = 0,$$

that is,

$$x = a + \beta \sqrt{-1} \text{ or } x = a - \beta \sqrt{-1} \dots \dots (4);$$

so that by making these substitutions we have, as in the former case,

$$\left[\frac{U}{V_1}\right] = [Ax] + B, \left[\frac{d \left\{ \dfrac{U}{V_1} - \&c. \right\}}{dx} \right] = 0, \left[\frac{d^2 \left\{ \dfrac{U}{V_1} - \&c. \right\}}{dx^2} \right] = 0, \&c.$$

each of these equations divides itself into two because of the two values of x (4), for which they subsist; hence we have as many equations as there are coefficients to be determined. It should be observed that in this second case the method of indeterminate coefficients, as explained in the text, is generally of easier application than that which we have just given, as the trouble of operating with the imaginary values (4) is avoided.

As an example of the foregoing processes let it be required to decompose the fraction

$$\frac{1}{x^8 + x^7 - x^4 - x^3}.$$

The denominator of this fraction is easily seen to be the same as

$$(x^4 - 1) \, x^3 \, (x + 1) = (x - 1) \, (x + 1)^2 \, x^3 \, (x^2 + 1)$$

so that we must assume

$$\frac{1}{x^8 + x^7 - x^4 - x^3} = \frac{A}{x - 1} + \frac{B}{(x + 1)^2} + \frac{C}{x + 1} + \frac{D}{x^3} + \frac{E}{x^2} +$$

$$\frac{F}{x} + \frac{Kx + L}{x^2 + 1}.$$

We shall first determine the numerator A, which, since $U = 1$ and $V_1 = (x + 1)^2 (x^2 + 1) x^3$, is

$$A = [\frac{1}{(x + 1)^2 (x^2 + 1) x^2}] = \frac{1}{8},$$

1 being substituted for x.

We shall next find B, C, which, since $V_1 = (x - 1) (x^2 + 1) x^3$ and $U = 1$, becomes

$$B = [\frac{U}{V_1}] = \frac{1}{4}, \quad C = [\frac{d \frac{U}{V_1}}{dx}] = \frac{9}{8},$$

-1 being substituted for x.

To determine D, E, F, we have $U = 1$, $V_1 = (x - 1) (x + 1)^2 (x^2 + 1)$,

$$D = [\frac{U}{V}] = -1, \quad E = [\frac{d \frac{U}{V_1}}{dx}] = 1, \quad F = \frac{1}{2} [\frac{d^2 \frac{U}{V_1}}{dx^2}] = -1,$$

0 being substituted for x.

Finally, to determine K, L, we have, since $U = 1$ and $V_1 = (x - 1) (x + 1)^2 x^3$,

$$[\frac{U}{V_1}] = [Kx] + L,$$

in which $\pm \sqrt{-1}$ being substituted for x gives the two equations

$$2K + 2L = 1, \quad K = L,$$

from which we get

$$K = \frac{1}{4}, \quad L = \frac{1}{4};$$

hence, the proposed fraction is decomposed into

$$\frac{1}{8(x - 1)} + \frac{1}{4(x + 1)^2} + \frac{9}{8(x + 1)} - \frac{1}{x^3} + \frac{1}{x^2} - \frac{1}{x} + \frac{x + 1}{4(x^2 + 1)}$$

3. If we had to integrate $\frac{dx}{x^n - 1}$ or $\frac{dx}{x^n + 1}$ we should have first to resolve the denominator into its quadratic factors, and this may be done by means of the decomposition of

$$x^{2m} - 2x^m \cos. \theta + 1,$$

already exhibited at page 32 of the Differential Calculus.

The cosine in the last or mth factor in this decomposition is obviously

$$\cos. \frac{\theta + 2(m-1)\pi}{m} = \cos. \left(2\pi + \frac{\theta - 2\pi}{m}\right) = \cos. \frac{\theta - 2\pi}{m}$$

the cosine in the factor preceding this is

$$\cos. \frac{\theta + 2(m-2)\pi}{m} = \cos. \left(2\pi + \frac{\theta - 4\pi}{m}\right) = \cos. \frac{\theta - 4\pi}{m},$$

and so on; so that the formula referred to may be written thus, by changing x into y and m into n, viz.

$$x^{2n} - 2x^n \cos. \theta + 1 = \left(x^2 - 2x \cos. \frac{\theta}{n} + 1\right)$$

$$\times \left(x^2 - 2x \cos. \frac{2\pi + \theta}{n} + 1\right)$$

$$\times \left(x^2 - 2x \cos. \frac{2\pi - \theta}{n} + 1\right)$$

$$\times \left(x^2 - 2x \cos. \frac{4\pi + \theta}{n} + 1\right)$$

$$\times \left(x^2 - 2x \cos. \frac{4\pi - \theta}{n} + 1\right) \times \&c. \text{ to } n \text{ terms}..(1).$$

Now it is easy to see here, that the last or nth factor is, *when n is even,*

$$x^2 - 2x \cos. \frac{n\pi + \theta}{n} = x^2 + 2x \cos. \frac{\theta}{n} + 1,$$

so that, in this case, the decomposition is

$$x^{2n} - 2x^n \cos. \theta + 1 = \left(x^2 - 2x \cos. \frac{\theta}{n} + 1\right)$$

$$\times \left(x^2 + 2x \cos. \frac{\theta}{n} + 1\right)$$

$$\times \left(x^2 - 2x \cos. \frac{2\pi + \theta}{n} + 1\right)$$

$$\times \left(x^2 - 2x \cos. \frac{2\pi - \theta}{n} + 1\right)$$

$$\times \left(x^2 - 2x \cos. \frac{4\pi + \theta}{n} + 1\right)$$

$$\times (x^2 - 2x \cos. \frac{4\pi - \theta}{n} + 1) \times \&c. \text{ to } n \text{ terms} .. (2).$$

Let us now suppose in each of these formulas (1) and (2) that $\theta = 0$, then cos. $\theta = 1$ as also cos. $\frac{\theta}{n}$, and, therefore, *when n is odd,*

$$x^{2n} - 2x^n + 1 = (x^2 - 2x + 1)$$

$$\times (x^2 - 2x \cos. \frac{2\pi}{n} + 1)^2$$

$$\times (x^2 - 2x \cos. \frac{4\pi}{n} + 1)^2 \times \&c. \text{ to } \frac{n+1}{2} \text{ terms} \ldots (3),$$

and *when n is even,*

$$x^{2n} - 2x^n + 1 = (x^2 - 2x + 1)$$

$$\times (x^2 + 2x + 1)$$

$$\times (x^2 - 2x \cos. \frac{2\pi}{n} + 1)^2$$

$$\times (x^2 - 2x \cos. \frac{4\pi}{n} + 1)^2 \times \&c. \text{ to } \frac{n+2}{2} \text{ terms} \ldots (4).$$

The formulas (3) and (4) immediately lead to the decomposition of $x^n - 1$ into its quadratic factors; for, by extracting the square root of each side of (3), we have, *when n is odd,*

$$x^n - 1 = (x - 1)$$

$$\times (x^2 - 2x \cos. \frac{2\pi}{n} + 1)$$

$$\times (x^2 - 2x \cos. \frac{4\pi}{n} + 1) \times \&c. \text{ to } \frac{n+1}{2} \text{ terms} \ldots (5),$$

and, by extracting the square root of each side of (4), we have *when n is even,*

$$x^n - 1 = (x^2 - 1)$$

$$\times (x^2 - 2x \cos. \frac{2\pi}{n} + 1)$$

$$\times (x^2 - 2x \cos. \frac{4\pi}{n} + 1) \times \&c. \text{ to } \frac{n}{2} \text{ terms} \ldots (6).$$

Having thus decomposed $x^n - 1$ into its quadratic factors, we may resolve $\dfrac{1}{x^n - 1}$ into its partial fractions, as follows:

Taking the logarithms of each side of (5), and then differentiating, we have, *when n is odd*,

$$\frac{nx^{n-1}}{x^n - 1} = \frac{1}{x - 1} + \frac{2x - 2\cos.\dfrac{2\pi}{n}}{x^2 - 2x\cos.\dfrac{2\pi}{n} + 1} +$$

$$\frac{2x - 2\cos.\dfrac{4\pi}{n}}{x^2 - 2x\cos.\dfrac{4\pi}{n} - 1} + \&c.\ \text{to}\ \frac{n+1}{2}\ \text{terms,}$$

or, multiplying by x,

$$\frac{nx^n}{x^n - 1} = \frac{x}{x - 1} + \frac{2x^2 - 2x\cos.\dfrac{2\pi}{n}}{x^2 - 2x\cos.\dfrac{2\pi}{n} + 1} +$$

$$\frac{2x^2 - 2x\cos.\dfrac{4\pi}{n}}{x^2 - 2x\cos.\dfrac{4\pi}{n} + 1} + \&c.\ \text{to}\ \frac{n+1}{n}\ \text{terms.}$$

Now, if we subtract n from the first side of this equation, and from each term on the second side 2, we shall subtract from the whole of this side $\dfrac{n+1}{2}$ times 2; that is $n+1$, so that, in order that the equation may still subsist, we must increase this remainder by 1, or, which is the same thing, the equation will subsist if we subtract n from the first side, and from the first term on the second side 1 only, while from every other term is taken 2; we shall thus have

$$\frac{n}{x^{n-1}} = \frac{1}{x - 1} - \frac{2 - 2x\cos.\dfrac{2\pi}{n}}{1 \quad 2x\cos.\dfrac{2\pi}{n} + x^2} -$$

$$\frac{2 - 2x\cos.\dfrac{4\pi}{n}}{1 - 2x\cos.\dfrac{4\pi}{n} + x^2} + \&c.\ \text{to}\ \frac{n+1}{2}\ \text{terms,}$$

and consequently

$$\frac{1}{x^{n}-1} = \frac{1}{n(x-1)} - \frac{2}{n} \left\{ \frac{1 - x \cos. \dfrac{2\pi}{n}}{1 - 2x \cos. \dfrac{2\pi}{n} + x^2} + \right.$$

$$\left. \frac{1 - x \cos. \dfrac{4\pi}{n}}{1 - 2x \cos. \dfrac{4\pi}{n} + x^2} + \&c. \text{ to } \frac{n-1}{2} \text{ terms} \right\}.$$

Hence the integration of

$$\frac{dx}{x^n - 1}$$

is, *when n is odd,* reduced to the integration of the several terms of the series

$$\frac{dx}{n(x-1)} - \frac{2dx}{n} \left\{ \frac{1 - ax}{1 - 2ax + x^2} + \frac{1 - bx}{1 - 2bx + x^2} + \right.$$

$$\left. \&c. \text{ to } \frac{n-1}{2} \right\} \text{ terms} \ldots \ldots (7),$$

which integration may be readily effected by the methods explained in the text.

Again, taking the logarithms of each member of the equation (6), and differentiating we have

$$\frac{nx^{n-1}}{x^n - 1} = \frac{2x}{x^2 - 1} + \frac{2x - 2 \cos. \dfrac{2\pi}{n}}{x^2 - 2x \cos. \dfrac{2\pi}{n} + 1} +$$

$$\frac{2x - 2 \cos. \dfrac{4\pi}{n}}{x^2 - 2x \cos. \dfrac{4\pi}{n} + 1} + \&c. \text{ to } \frac{n}{2} \text{ terms.}$$

or, multiplying by x,

$$\frac{nx^n}{x^n - 1} = \frac{2x^2}{x^2 - 1} + \frac{2x^2 - 2x \cos. \dfrac{2\pi}{n}}{x^2 - 2x \cos. \dfrac{2\pi}{n} + 1} +$$

$$\frac{2x^2 - 2x \cos. \dfrac{4\pi}{n}}{x^2 - 2x \cos. \dfrac{4\pi}{n} + 1} + \&c. \text{ to } \frac{n}{2} \text{ terms.}$$

Subtract n from the first member of this equation, and as there are $\dfrac{n}{2}$ terms in the second member, subtract 2 from each, the result is

$$\frac{n}{x^n - 1} = \frac{2}{x^2 - 1} - \frac{2 - 2x \cos. \dfrac{2\pi}{n}}{1 - 2x \cos. \dfrac{2\pi}{n} + x^2} -$$

$$\frac{2 - 2x \cos. \dfrac{4\pi}{n}}{1 - 2x \cos. \dfrac{4\pi}{n} + x^2} + \&c. \text{ to } \frac{n}{2} \text{ terms,}$$

consequently

$$\frac{1}{x^n - 1} = \frac{2}{n (x^2 - 1)} - \frac{2}{n} \left\{ \frac{1 - x \cos. \dfrac{2\pi}{n}}{1 - 2x \cos. \dfrac{2\pi}{n} + x^2} + \right.$$

$$\left. \frac{1 - x \cos. \dfrac{4\pi}{n}}{1 - 2x \cos. \dfrac{4\pi}{n} + x^2} + \&c. \text{ to } \frac{n - 2}{2} \text{ terms} \right\}.$$

Hence the integration of

$$\frac{dx}{x^n - 1}$$

is *when n is even* reduced to the integration of the several terms of the series

$$\frac{2dx}{n (x^2 - 1)} - \frac{2dx}{n} \left\{ \frac{1 - ax}{1 - 2ax + x^2} + \frac{1 - bx}{1 - 2bx + x^2} + \right.$$

$$\left. \&c. \text{ to } \frac{n - 2}{2} \text{ terms} \right\} \ldots . (8).$$

We shall now give an example of each of the formulas (7) and (8):

1. To determine the integral of

$$\frac{dx}{x^3 - 1}$$

By the formula (7) we have

$$\int \frac{dx}{x^3 - 1} = \frac{1}{3} \int \frac{dx}{x - 1} - \frac{2}{3} \int \frac{(1 - ax)\,dx}{1 - 2ax + a^2},$$

the first integral in the second member is

$$\frac{1}{3} \log. (x - 1),$$

and, by putting the denominator of the remaining integral under the form

$$(x - a)^2 - a^2 + 1,$$

and then substituting z for $x - a$, we have

$$\int \frac{(1 - ax)\,dx}{1 - 2ax + x^2} = (1 - a^2) \int \frac{dz}{z^2 - a^2 + 1} - a \int \frac{z\,dz}{z^2 - a^2 + 1}$$

$$= \frac{1 - a^2}{\sqrt{1 - a^2}} \tan.^{-1} \frac{1}{\sqrt{1 - a^2}} z - \frac{a}{2} \log. (z^2 - a^2 + 1);$$

hence, substituting for a and z their values

$$a = \cos. \frac{2\pi}{3}, \; z = x - a,$$

we have, for the required integral,

$$\int \frac{dx}{x^3 - 1} =$$

$$C + \frac{1}{3} \log. (x - 1) + \frac{1}{3} \cos. \frac{2\pi}{3} \log. (1 - 2x \cos. \frac{2\pi}{3} + x^2) -$$

$$\frac{2}{3} \sin. \frac{2\pi}{3} \tan.^{-1} \frac{x - \cos. \dfrac{2\pi}{3}}{\sin. \dfrac{2\pi}{3}}.$$

If this integral ought to vanish when $x = 0$, then the correction is

$$C = \frac{2}{3} \sin. \frac{2\pi}{3} \tan.^{-1} \frac{- \cos. \dfrac{2\pi}{3}}{\sin. \dfrac{2\pi}{3}}$$

$$= -\frac{2}{3} \sin. \frac{2\pi}{3} \tan.^{-1} \frac{\cos. \dfrac{2\pi}{3}}{\sin. \dfrac{2\pi}{3}}$$

and consequently the last term in the above expression, when corrected, becomes, in consequence of the property

$$\tan.(A + B) = \frac{\tan. A + \tan. B}{1 - \tan. A \tan. B},$$

$$-\frac{2}{3} \sin. \frac{2\pi}{3} \tan.^{-1} \frac{x \sin. \dfrac{2\pi}{3}}{1 - x \cos. \dfrac{2\pi}{3}},$$

2. If the integral of

$$\frac{dx}{x^4 - 1}$$

is required, then, by the application of the formula (8), we find

$$\int \frac{dx}{x^4 - 1} =$$

$$-\frac{1}{4} \log. \frac{1 + x}{1 + x} + \frac{1}{4} \cos. \frac{\pi}{2} \log. (1 - 2x \cos. \frac{\pi}{2} + x^2) -$$

$$\frac{1}{2} \sin. \frac{\pi}{2} \tan.^{-1} \frac{x \sin. \dfrac{\pi}{2}}{1 - x \cos. \dfrac{\pi}{2}},$$

the integral being corrected as in the preceding example. Or, since $\cos. \frac{\pi}{2} = 0$, and $\sin. \frac{\pi}{2} = 1$,

$$\int \frac{dx}{x^4 - 1} = -\frac{1}{4} \log. \frac{1 + x}{1 - x} - \frac{1}{2} \tan.^{-1} x.$$

Let it now be required to decompose $x^n + 1$. For this purpose, put $\theta = \pi$, in the formula (1), and it becomes

$$x^{2n} + 2x^n + 1 = (x^2 - 2x \cos. \frac{\pi}{n} + 1)^2$$

$$\times (x^2 - 2x \cos. \frac{3\pi}{n} + 1)^2$$

$$\times (x^2 - 2x \cos. \frac{5\pi}{n} + 1)^2 \times \&c. \text{ to } \frac{n}{2} \text{ terms.}$$

Hence, by extracting the square root, we have, *when n is even,*

$$x^n + 1 = (x^2 - 2x \cos. \frac{\pi}{n} + 1)$$

$$\times (x^2 - 2x \cos. \frac{3\pi}{n} + 1)$$

$$\times (x^2 - 2x \cos. \frac{5\pi}{n} + 1)$$

and in like manner by putting $\theta = \pi$, in the formula (2), and extracting the square root, we have, *when n is odd,*

$$x^n + 1 = (x + 1)$$

$$\times (x^2 - 2x \cos. \frac{\pi}{n} + 1)$$

$$\times (x^2 - 2x \cos. \frac{3\pi}{n} + 1) \times \&c. \text{ to } \frac{n+1}{2} \text{ terms;}$$

hence, proceeding exactly as with $x^n - 1$, in the respective cases of n even and n odd, we find, *when n is even,*

$$\frac{dx}{x^n + 1} = \frac{2dx}{n} \left\{ \frac{1 - x \cos. \frac{\pi}{n}}{1 - 2x \cos. \frac{\pi}{n} + x^2} + \frac{1 - x \cos. \frac{3\pi}{n}}{1 - 2x \cos. \frac{3\pi}{n} + x^2} + \right.$$

$$+ \&c. \text{ to } \frac{n}{2} \text{ terms}$$

$$= \frac{2dx}{n} \left\{ \frac{1 - ax}{1 - 2ax + x^2} + \frac{1 - bx}{1 - 2bx + x^2} + \right.$$

$$\&c. \text{ to } \frac{n}{2} \text{ terms} \right\} \ldots \ldots (9)$$

and when *n* is odd,

$$\frac{dx}{x^n + 1} = \frac{dx}{n(1+x)} + \frac{2dx}{n} \left\{ \frac{1 - x \cos. \dfrac{\pi}{n}}{1 - x \cos. \dfrac{\pi}{n} + x^2} + \right.$$

$$\frac{1 - x \cos. \dfrac{3\pi}{n}}{1 - 2x \cos. \dfrac{3\pi}{n} + x^2} + \&c. \text{ to } \frac{n-1}{2} \text{ terms}\}$$

$$= \frac{dx}{n(1+x)} + \frac{2dx}{n} \left\{ \frac{1 - ax}{1 - 2ax + x^2} + \right.$$

$$\frac{1 - bx}{1 - 2bx + x^2} + \&c. \text{ to } \frac{n-1}{2} \text{ terms}\} \ \ldots \ (10).$$

If we apply this formula to the example

$$\frac{dx}{x^3 + 1},$$

we find for the integral

$$\int \frac{dx}{x^3 + 1} =$$

$$\frac{1}{3} \log. (1 + x) - \frac{1}{3} \cos. \frac{\pi}{3} \log. (1 - 2x \cos. \frac{\pi}{3} + x^2) +$$

$$\frac{2}{3} \sin. \frac{\pi}{3} \tan.^{-1} \frac{x \sin. \dfrac{\pi}{3}}{1 - x \cos. \dfrac{\pi}{3}},$$

the same correction being introduced as in the former examples. Or, since cos. $60° = \frac{1}{2}$ and sin. $60° = \frac{1}{2} \sqrt{3}$

$$\therefore \int \frac{dx}{x^3 + 1} = \frac{1}{3} \log. \frac{1 + x}{\sqrt{1 - x + x^2}} + \frac{1}{\sqrt{3}} \tan.^{-1} \frac{x \sqrt{3}}{2 - x}.$$

For the decomposition and integration of other forms, the student may consult *Jephson's Fluxional Calculus*, vol. 2, and *Simpson's Fluxions*, vol. 2.

D d

Note (B), *page* 71.

Development of $\sin.^m x$ *and* $\cos.^m x$.

Put

$$\left. \begin{array}{l} \cos. x + \sin. x \sqrt{-1} = u \\ \cos. x + \sin. x \sqrt{-1} = v \end{array} \right\} \ \ldots . (1),$$

then (*Diff. Calc.* p. 30,)

$$\left. \begin{array}{l} \cos. mx + \sin. mx \sqrt{-1} = u^m \\ \cos. mx + \sin. mx \sqrt{-1} = v^m \end{array} \right\} \ \ldots . (2),$$

and consequently

$$u^m + v^m = 2 \cos. mx, \ u^m v^m = 1 \ \ldots . (3).$$

Now by adding together the equations (1), we get

$$\cos. x = \frac{1}{2}(u + v),$$

and therefore

$$\cos.^m x = \frac{1}{2^m}(u + v)^m = \frac{1}{2^m}(v + u)^m;$$

hence, by the binomial theorem,

$$\cos.^m x = \frac{1}{2^m}\left\{ u^m + m u^{m-1} v + \frac{m(m-1)}{2} u^{m-2} v^2 + \&c. \right\}.$$

or

$$\cos.^m x = \frac{1}{2^m}\left\{ v^m + m v^{m-1} u + \frac{m(m-1)}{2} v^{m-2} u^2 + \&c. \right\},$$

adding together these equations, we have

$$\cos.^m x = \frac{1}{2^{m+1}}\left\{ u^m + v^m + muv(u^{m-2} + v^{m-2}) + \right.$$

$$\frac{m(m-1)}{2} u^2 v^2 (u^{m-4} + v^{m-4}) + \&c. \right\}$$

But from (3)

$$u^m + v^m = 2 \cos. mx \qquad\qquad uv = 1$$
$$u^{m-2} + v^{m-2} = 2 \cos. (m-2) x \qquad u^2 v^2 = 1$$
$$u^m{-4} + v^{m-4} = 2 \cos. (m-4) x \qquad u^4 v^4 = 1$$
$$\&\text{c.} \qquad \&\text{c.} \qquad\qquad \&\text{c.} \quad \&\text{c.}$$

hence the development of $\cos.^m x$ becomes

$$\cos.^m x = \frac{1}{2^m} \{\cos. mx + m \cos. (m-2) x +$$
$$\frac{m(m-1)}{2} \cos. (m-4) x + \&\text{c.}\} \ \dots \ (A.),$$

and by putting m equal to 2, 3, 4, &c. and recollecting that $\cos. - \varphi = \cos \varphi$, we shall obtain the values of $\cos.^2 x$, $\cos.^3 x$, $\cos.^4 x$, &c. in the text.

Let us now seek the development of $\sin.^m x$; for this purpose we must take the difference, instead of the sum, of the equations (1); we thus have

$$2 \sin. x \sqrt{-1} = u - v \ \therefore \ \sin. x = \frac{u - v}{2 \sqrt{-1}},$$

and consequently

$$\sin.^m x = \frac{(u - v)^m}{(2\sqrt{-1})^m}.$$

1. Let m be even, then (*Algebra*, p. 149,)

$$(u - v)^m = (v - u)^m;$$

hence, developing the two equations

$$\sin.^m x = \frac{1}{(2\sqrt{-1})^m} (u - v)^m, \ \sin.^m x = \frac{1}{(2\sqrt{-1})^m} (v - u)^m,$$

and adding the results, we get

$$2 \sin.^m x = \frac{1}{(2\sqrt{-1})^m} \{u^m + v^m - muv (u^{m-2} + v^{m-2}) +$$
$$\frac{m(m-1)}{2} u^2 v^2 (u^{m-4} + v^{m-4}) - \&\text{c.}\}$$

and making the same substitutions as before, in virtue of (3), we have, since m is even, and therefore $(\sqrt{-1})^m = \pm 1$,

$$\sin.^m x = \pm \frac{1}{2^m} \{\cos. mx - m \cos. (m-2)x +$$

$$\frac{m(m-1)}{2} \cos. (m-4)x - \&c.\} \ldots (B).$$

2. Let m be odd, then

$$(u-v)^m = (-1)^m (v-u)^m = -(v-u)^m,$$

therefore

$$\sin.^m x = \frac{(u-v)^m}{(2\sqrt{-1})^m}, \sin.^m x = -\frac{(v-u)^m}{(2\sqrt{-1})^m},$$

and developing $(u-v)^m$, and $(v-u)^m$, as before, we get

$$2\sin.^m x = \frac{1}{(2\sqrt{-1})^m} \{u^m - v^m - muv(u^{m-2} - v^{m-2}) +$$

$$\frac{m(m-1)}{2} u^2 v^2 (u^{m-4} - x^{m-4}) - \&c.\}$$

But from equations (2)

$$u^m - v^m + 2\sin. mx \sqrt{-1}, u^m v^m = 1,$$

and in virtue of these equations the foregoing development becomes, since $(\sqrt{-1})^{m-1} = \pm 1$,

$$\sin.^m x = \pm \frac{1}{2^m} \{\sin. mx - m \sin. (m-2)x +$$

$$\frac{m(m-1)}{2} \sin. (m-4)x - \&c.\} \ldots (C).$$

It must be observed, that in the development (B), the lower sign is to be employed, when m is either of the numbers 2, 6, 10, &c. and the upper sign, when m is either of the numbers 4, 8, 12, &c. Also in the development (C), the lower sign is to be used, when $m - 1$ is one of the numbers in the first series, and the upper sign, when it is one of the numbers in the second series.

Note (C), *page* 142.

In order to show that every enveloping surface must be greater than the surface enveloped, it must be first established that every surface, curved or polygonal, which is subtended by a plane exceeds that plane. This will be obvious, from considering that if through any point on the curve, or polygonal surface, a plane be drawn, intersecting both the surface and the subtending plane, the curve section will always exceed the rectilinear section, whatever be the direction of the intersecting plane; and that, therefore, the locus of the curve sections, that is, the curve surface, must exceed the locus of the rectilinear sections, or the subtending plane. This being admitted, let us conceive any two surfaces, one enveloping the other; then, as there is necessarily some space between them, we may cut off by a plane a portion of the enveloping surface without touching the surface enveloped; if then this plane supply the place of the portion cut off, the enveloping surface thus modified will be less than before, and the space between it and the enveloped surface will be diminished. Again, let the intermediate space be still further diminished, by cutting off another portion of the enveloping surface, and let this process be continued; then it is obvious, that since at every operation we not only diminish the enveloping surface, but also the space between the two, we in fact approach nearer and nearer to coincidence to the enveloped surface, as the enveloping surface diminishes; consequently this latter must have been originally greater than the former.

Note (D), *page* 231.

We have remarked in the text that there does not exist any singular solution when the arbitrary constant c enters into the complete primitive only in the first power. This, however, is contrary to the doctrine of most analytical writers, who, in cases of this kind, reason as follows:

"When c rises only to the first degree in the primitive, this is of the form

$$u = Ac + P = 0 \ . \ . \ . \ . \ (1),$$

where A and P are functions of x, y, which do not contain c. First, suppose A not to be a factor of P; then, since $c = -\dfrac{P}{A}$, from $\dfrac{du}{dc} = A = 0$, there results $c = \infty$; which gives a particular solution, viz. that case of the primitive in which the arbitrary constant is supposed to be infinite.

"Next, let A be a factor of P; then, since the proposed differential equation is

$$P\,(dA) - A\,(dP) = 0,*$$

$A = 0$ must necessarily be a solution, and to determine whether it is a singular solution, eliminate either x or y from $A = 0$ and the primitive; and it is a singular solution or not, according as the resulting value of c is variable or constant."

It would appear, then, from this reasoning, that $A = 0$ might be a singular solution, provided the complete integral (1) were divisible by A; but in such a case the solution $A = 0$ would always be necessarily comprised in every *particular* solution (1): this solution cannot, therefore, with propriety be considered a singular solution, for it is the character of a singular solution not to be comprised in any particular solution.

---

* For the immediate differential of

$$Ac + P = 0 \ . \ . \quad . \ (1)$$

is

$$c\,(dA) + dP) = 0 \therefore c = -\frac{(dP)}{(dA)} \therefore (1), P - A\frac{(dP)}{(dA)}$$

$$\therefore P\,(dA) - A\,(dP) = 0.$$

THE END.

Made in the USA
Monee, IL
07 July 2026

56551657R00177